ODD LOTS

ODD LOTS

Ideas, Humor, Parody and Memoir
from a Contrarian Optimist

Jeff Duntemann

Copperwood Press • Scottsdale, Arizona
2021

ODD LOTS

ISBN 10: 1-932084-15-0
ISBN 13: 978-1-932084-15-3

Cover:

An arrangement of odd items from the author's "Box of No Return," including (among other things) various name badges, a packet of gold leaf, a cow's molar, a pair of Hewlett-Packard high-power op amps, some junior ROTC brass, Comdex vendor buttons, a Matchbox toy bus, some Japanese paper money, a 2708 EPROM chip, a sealing wax candle, a shell casing from the 21-gun salute given at his father's funeral, some old keys to forgotten locks, a piece of pyrite, and a first-aid kit built into an old medium-format film can as a project in Boy Scouts.

Photo by the author.

For all of my former colleagues at
The Coriolis Group 1989-2002
without whom a lot of this material would not exist.

TABLE OF CONTENTS

INTRODUCTION

I put this book together for my fans. I've been a writer for a long time, and some of my fans have been with me for decades. I was writing stories for fun and passing them around among my close friends since seventh or eighth grade. I wrote a few things for my high school literary magazine, and later my college newspaper. I sold my first SF story and my first nonfiction article into professional markets in the same year: 1973, though neither appeared in print until 1974. I've written all or part of twenty-odd technical books since 1985. I founded and published a technical magazine for programmers from 1990-2000, and wrote something (often several somethings) in every issue. I independently invented blogging in 1998 (more on which later) and have been blogging ever since. I've been self-publishing my science fiction and fantasy since 2010. I've been on Twitter and Facebook for years (and most recently, Facebook lookalike MeWe) but in my view that barely counts as writing. I do it mostly to promote my self-published fiction.

That's a lot of words. Some of it—almost entirely the very early stuff—has been lost. That's ok; some things richly deserve to be lost. I have over 100,000 words down on my memoirs, and although they're not lost, I didn't write them to be published. (I may share them with close friends who express interest, thereby completing the circle.) Mostly I wrote them to record elements of my life that I remember *now*—but may not remember in twenty or thirty years.

A great deal of my technical writing is now obsolete. That's to be expected, given that I wrote a lot of it thirty-odd years ago. Time was, there were four or five programming languages that did almost anything that needed doing. Now we have dozens, with new ones popping up all the time, like mushrooms after a hard rain—and the ones I did most of my work in, like FORTRAN, BASIC, and Pascal, are considered, at best…quaint. I've decided to exclude all of that in this book. But here and there amidst that huge pile of writing are short items that are interesting, peculiar, funny, prophetic, or all of the above. People sometimes ask me for an editorial or an idea piece they remember from *PC Techniques/Visual*

Developer. Some of those are online; most are not. The idea here is to provide a collection of essays, idea pieces, song and poetry parodies, and blog entries of lasting interest—plus a couple of odd things that fit in no convenient category. They are well and truly Odd Lots. There are a lot of them. This book contains most of them that are still worth reading. After many of the pieces I've added a little verbiage putting them into context, especially for people who have not read them before.

Because this book will come as close as I'll get to publishing an autobiography, I've included a little memoir where it makes sense. My life, mercifully, has been smooth and generally lacking in sharp edges. People who want to read about celebrity sexual escapades or drug habits will have to keep looking. People who believe in the future—and in the power of technology to improve that future—will find themselves very much at home. So grab a bowl of Cheetos and an adult beverage of your choice, lean back in a comfy chair, and allow me to share some ideas, perspectives, and encouragements that have been piling up during the last (almost) fifty years.

PART 1. EDITORIALS AND IDEA PIECES

My friend Keith Weiskamp and I created a publishing company in late 1989. Our first product was a general-topics magazine for programmers, *PC Techniques*. In 1994 we began publishing books, and in 1996 we changed the name of the magazine to *Visual Developer*. The magazine ran for sixty issues—ten years—and folded in early 2000. I wrote an editorial in every issue, and in most issues I wrote an idea or humor piece for the last page. In this section I'm publishing a selection of those essays and editorials that first appeared in our magazine.

Many if not most of my essays from the magazine were tied to their times and haven't aged well. Remember that even the newest of them are now over twenty years old, and the oldest over thirty. I've left those online—or, in some cases, in the paper magazines. I've written short afterwords at the ends of most of the pieces included here to put them in context, and sometimes to point up how something I predicted back in the '90s came to pass—or didn't.

Some of those last-page articles were programmer humor and parody, which I've saved for Part 3.

The All-Volunteer Virtual Encyclopedia of Absolutely Everything!

From *PC Techniques Magazine #27*, June/July 1994

Crazy world, ain't it? Yes indeedy, and we let a lot of that craziness go completely to waste. A guy I know used to breed Japanese quail and sell pickled quail eggs for a dollar apiece to lonely Japanese expatriates on the West Coast. He knows a lot about Japanese quail. (For example: Put them in *tall* coops—loud noises make them jump straight up, with sufficient force to break their own necks on a low roof. One thunderstorm and your whole flock might have to be quick-frozen for TV dinners.) If you wanted to tap this ancient and venerable quail wisdom, where would you go?

Nowhere is where. The time you'd spend looking would probably be worth far more than the found knowledge, unless you were as quail-crazy as Don. But within our grasp (if we ever do get an Information Superhighway) will soon be the means to create an All-Volunteer Virtual Encyclopedia of Absolutely Everything, pickled quail eggs pointedly included.

Face it: A hireling at an encyclopedia company cannot write about quail husbandry with the verve and nuttiness of someone to whom it is a total lifestyle obsession. The same goes for topics like pre-World War II microwave power tubes, Pez dispensers of the Sixties, or the Peppermint Trolley Company and its Greatest Hit. So let's envision an NII node with a few spare gigabytes, and set up a master index to what may someday become the elucidation of All Human Knowledge.

Someone (and I volunteer) would write a style guide for encyclopedia entries. One *can* know too much about Pez dispensers; 2,000 words is probably enough. That done, word would go out on the Net of Nets: If you care passionately about something and consider yourself an expert, write it up, put it somewhere where it can be found, and mail that address to the Encyclopedia. The Encyclopedia itself doesn't contain the entries, only an index line and where to do the ftp. If the NII is fast enough to handle teleconferencing and multimedia, it will be fast enough to do an ftp on a 2,000-word article in a couple of minutes or less.

The intellectual richness of the Internet-cum-NII community is mind-boggling, and will soon number in the tens of millions. If even a fraction of those people contribute an entry or two, the Encyclopedia will soon sport hundreds of thousands of entries, with ten or twenty thousand new ones coming in every year. Nor will all those be about the life and times of Edward VII, or the agricultural economy of Gambia. Most of them will be the sorts of things that simply can't be known anywhere else, which is precisely the sort of knowledge that dies with its owners, because its owners have no way to pass it on.

Organizations with gigabytes to spare might offer Encyclopedia-caching services; that is, they would ftp the most requested entries of interest to specialty audiences onto fast disk so that like-minded folks could browse it all in real-time. Astronomers might subscribe to one such cache, where Otto Struve would be present, and Edward VII absent, with Edward and his namefellows replacing Otto & Co. in a cache for Royalty weenies. You can browse the Bichon Frise online stud books in the Frise Rampant cache, along with biographies of every Bichon of note, including Mr. Byte—and if you still wanted to look up poor Edward, well, the cache would dutifully send for him, and cache him with the canines. If enough people asked for him, the least recently used (LRU) cache would simply make him a regular, whether he was a Bichon, owned one, or simply resembled one.

Guys like me, who read the dictionary for fun, would probably not find a cache of sufficient breadth to be satisfying. Hey, I'd wait a few minutes for the authorized history of the 829B power pentode, and while I was waiting, I'd simply read the life and times of Big Daddy Don Garlitz or catch up on the current state of instrumental transcommunication.

Quail you want? *Quail* we'll give you!

I doubt I need to say this 25-odd years later, but we have achieved the All-Volunteer Virtual Encyclopedia of Absolutely Everything, and its name is Wikipedia. I wrote this piece in February, 1994. The Web was still in its infancy then, and I was thinking in terms of FTP and the National Information Infrastructure (NII), better known back then (though by now forgotten) as the Information Superhighway. Today we have HTTP and ubiquitous broadband.

I'm hardly the first person to have imagined a globe-spanning encyclopedia. H. G. Wells described what he called The World Brain in 1937.

It's fun to see that something I had predicted came to pass more or less as I had hoped it would. Wikipedia has its problems, like a pathological obsession with petty rules on nebulous factors like "notability." There are other ways to skin this cat, including InfoGalactic, which is an independent MediaWiki-based fork of Wikipedia that doesn't care half as much about notability as it does about the accuracy of its material. *That's* the future I was envisioning—but in truth, didn't dare hope we would achieve.

Still, of all the SFnal predictions I've ever made, this is probably the best.

Jiminy!

From *PC Techniques Magazine* #14, June/July 1992

My computer is pinned to my lapel. Generically, his kind are called *jiminies*, after that cricket guy. Picture a 19th-centuriy pine-box coffin-shape made of black plastic, perhaps 4 inches in length. Two dull red lenses are set like eyes at the widest part. One is a broadbeam single-wavelength infrared transmitter, the other a receiver. They operate at one gigabit per second. In the top of the coffin is a combined speaker-microphone.

This is Ragpicker. He talks to me through the speaker mike, from his perch on my lapel. He talks to the rest of the universe on infrared. He is both CPU and main memory, nonvolatile and extremely rugged. He contains about 64 terabytes of storage, divided among 256 parallel processors capable of operating independently or together.

In my briefcase is a 2000 dpi color pad, 9 inches by 12 inches, and a half inch thick. The pad has "eyes" like Ragpicker, but very little computing power. When I pick up the stylus the backlight comes on, and Ragpicker talks to me through the stylus and the pad. If I sketch a chart or a circuit. Ragpicker remembers it, and nothing remains in the pad to be stolen or lost.

At home I have similar pads here and there, with a large one and a good keyboard to one side of my desk. When I sit down at them, they come alive at Ragpicker's command. But he is the computer—they're only dumb peripherals. No matter where I go, to work or to friends' homes, the peripherals all obey Ragpicker, and all my favorite work habits—Ragpicker's carefully-learned habits—are available wherever I find a keyboard or pad/screen.

Ragpicker listens carefully, and speaks well. He isn't terribly imaginative, but that's OK—ideas are *my* job. He has two special skills: Knowing where information is stored, and judging what data is relevant to a query. That's crucial, since an information density of 16 terabytes per $2 cubic centimeter means that nothing is ever forgotten, no matter how trivial.

Where is the information? Mainly in electronic public libraries, or in for-profit information brokerages on common-carrier fiber networks. But there is also something called Infranet—the triumph of the old Fidonet idea, in that every computer is a node, and every node connects automatically to any other node to the limit (about 100 feet) of their infrared eyes.

If I need to know something, I define a query verbally until Ragpicker and I agree that it's rational and closed-ended enough to be useful. Then Ragpicker creates a tightly-focused query subset of himself and sends it out via infrared to find some answers for me. Such query objects, called *ghosts*, are executed at each node they enter. And if what they seek isn't in a given node, they find out quickly and transmit themselves on to the next. Information isn't always free, but Ragpicker has a budget and knows how to dicker. (Ghosts can carry video or audio messages as well.)

Most of the time, a computer's infrared eyes are idle, and that idle time is used to pass ghosts back and forth, from hand to hand, around your office or even up and down the crowd on a noon-time boulevard. The secret of the Infranet is that the Net is civilization itself.

Ninety percent of all people carry a node around on their lapels or built into their hats or the frames of their glasses—and every node passes ghosts along peaceably, allowing them to ask questions as they pass through. No one is required to answer—but almost everyone tries to help.

All fixed-position nodes are required by law to know and report their world coordinates when asked. So nobody is ever lost, and no ghost ever forgets how to return home. Minutes or hours later, Ragpicker is handed his ghost back from a nearby fiberport or even some passerby, and Ragpicker reports what the ghost has found. Usually, it's an answer, and possibly additional "surprise" answers to standing queries that the ghost has asked of total strangers as it passed by. Perhaps a guy with that rare Leslie Gore album I've been looking for, or somebody who wants to buy the Collins 75A4 I'm selling.

All the world's a bulletin board now. Ragpicker allows me to write, draw, calculate, correlate and analyze, to communicate and just mess around.

This is where the arrows converge: Nothing is forgotten; everything is connected; and the UI is Structured English, written or spoken. I expect to see it before I die. It will be an interesting world indeed.

I still think this prediction will turn out more or less as conceived in 1992. We have smartphones with more smarts than the entire Apollo program had—which got us to the Moon not once but several times. (Alas, not enough.) The phones are bigger than Ragpicker because in 1992, I didn't imagine that small but usefully dense color displays would be built in my lifetime. (This was one of my biggest blind spots when pondering the future of computing.) If you would be content with a phone that talked and listened to voice commands, well, we're there already and have been for some years. Still, keep in mind that if we wanted jiminies right now (2021) we could have them. What I didn't imagine (being more of a reader than a watcher) was binging on Star Trek TOS on your pocket telephone.

I need to add here that I didn't get this idea in 1992. The notion of "jiminies" goes back to an aborted SF novel I was tinkering with in 1983 and abandoned around 1990. The notion of semi-intelligent queries called "ghosts" goes back even further, to a never-published story I wrote at the Clarion workshop in 1973, before personal computers even existed. The ghosts from that story ran in a cylinder roughly the size of a thermos bottle, and zipped around the world to a small number of gigantic memory blocks created inside climate-controlled buildings 1000 feet high. This was definitely the George O. Smith mainframe vision of a global computer network. Not surprising: I learned computing on an IBM System/360, and personal computers did not appear until I had already graduated from college.

Computers, Like Dirt

From *PC Techniques Magazine* #15, August/September 1992

In this space last issue, I described my vision of computers with most of the power of a Cray-2—that you clip to your lapel. Several correspondents chided me for failure of imagination. "What about nanocomputers?" they asked. Oh, yeah. Just what we need. Computers that are too small to see.

Don't laugh. I'm not making this up. There's an intriguing book called *Engines of Creation* by K. Eric Drexler (Doubleday, 1986) that brought nanotechnology to public attention. I recommend the book despite its breathlessness, its undefended materialist view of the human mind, and lots of hand-waving in critical places.

But hey, that was 1986, and the gist of it is this: We are rapidly developing the ability to mechanically manipulate matter *one atom at a time*. That is, we will soon be able to pick up a silicon atom, and *slam!* stick it to the side of a beryllium atom, creating silicon beryllide, which does not exist in nature and could be anything from a new rat poison to an aphrodisiac. IBM recently hauled a bunch of xenon atoms across a copper plate, and dropped them accurately enough to spell—need we ask?—"IBM," in a 5x5 bitmapped font, where each bit was xenon atom. Think of *that* the next time you complain about small print.

Now, this is a cool trick, but it's really the equivalent of making mud pies when what you want is high-tech ceramics. A whole body of research is scouting the way to make computers out of individual atoms. Electronics is out, since individual electrons are hard to find, much less control. Instead, we're working toward a realization of Babbage's mechanical Difference Engine, only with rods and cams made of mere handfuls of atoms. The rods slide within sleeves, (lubricated by heaven-knows-what) such that a rod at one end of its travel indicates a 1, and a rod at the other end of its travel indicates a 0. Bumps on the rods allow them to "switch" other rods, create logic gates, etc. Because the rods are exceedingly small and have vanishingly little mass, the switching happens at the equivalent of gigahertz clock rates. Mass storage would be on "tapes" of material resembling DNA, with "readers" scooting along the strands, executing subroutines and updating data.

As messy as it sounds, a very serious computer made this way would be about the size of a bacterium, and a supercomputer would be no larger than a dust panicle. Is this possible? Sure. A virus is a sort of invasive molecular microcontroller, and a bacterium could be thought of as a molecular supercomputer robot with bad software. There is talk of slipping nanocomputer controllers into bacteria and injecting millions of them into the bloodstreams of AIDS patients, to follow HIV right into the nuclei of infected cells, and tear the viruses out by the roots.

It's a whole new class of problems. I've hurt my back carrying computers and cut my fingers repairing computers but I've never had to worry about *inhaling* computers before. And this implies that nanocomputers will probably become components inside mechanisms that we can see; the jiminy on your lapel, for example, will not merely store a transcript of your days' activities, but *all human knowledge*—and still be mostly empty space.

"Naked" nanocomputers will certainly have their uses. Imagine a device the size of a grain of dust (one cubic mil) with one face an image sensor. The rest of the device is a bucket-brigade image storage system that stores millions of images, clocking in a new one every second, or minute, or hour, in effect taking "movies" lasting hours, days, or years. Now imagine untold trillions of these little camcorders released into the environment and carried by the winds to every corner of the earth.

No matter where you go, the very dirt on the street is taking your picture. Even in your own home, the dust that Mr. Byte tracks in watches your arguments, your deceits—literally your every move, at 5000 X 5000 resolution. Want to solve a crime? Go back to the murder site and dig a thimbleful of dust from several points, and you've got millions of movies of the murder as it happened. Rob a bank and the dirt on the floor convicts you. Cleanliness is statistical; no matter how clean an environment, the dust is down there somewhere.

Nanocomputers could make it impossible to commit crimes of any sort undetected—or to keep secrets of any kind at all. Virtue imposed by the dust on the wind—how's that for an endpoint to the evolution of computing?

I thought this was a bit of a stretch when I first wrote it, in 1992. But hey, I'm a science fiction writer! Stretching reality is what we do for a living! Nano was the hot ticket in the midlate '90s. I wrote my first novel about it. I assumed we'd have nanocomputers by 2020—simple proof-of-concept ones, at least. No joy. We've still got some ore to mine in 3-D fabrication of conventional microcircuitry. Once that ore gives out, I suspect nano-computers will start to happen. 2040? We'll see.

I didn't think of it in those terms in 1992, but what this essay is really about is the idea of privacy, and how it may ultimately become impossible, irrespective of our attempts to legislate privacy into our society. Smart dirt? We've seen weirder things in this increasingly weird era.

Pay Them Forward

From *PC Techniques Magazine* #14, June/July 1992

Back in Rochester, New York in 1983, I met an extraordinarily bright eighth grader who was writing animated video games—good ones—in interpreted BASICA on his father's PC. He had that old familiar hunger everywhere about him, and I knew that he would become a major talent in a few years.

So I started giving him old copies of books and software, even after I left Rochester to join *PC Tech Journal* in Baltimore. I critiqued his software. I played his video games. I told him to keep at it. In short, I took him seriously; forgetting, at the time, what it means to an awkward 13-year-old to be taken seriously by an adult.

A year or two later, he wrote me a letter thanking me for the software and the help. In the process, he used a remarkable phrase: He said that he could never pay me back, so instead he would pay me forward, and someday pass along his encouragement and old software to some kid whom he thought might have potential.

I won't embarrass him by naming him, since he now writes for this magazine, and is as private a person as I am an insufferable extrovert. But in three words he took me back a long, long way—and I remembered being 13 myself, and I remembered Uncle Louie.

He was the family black sheep; the unconventional, unmarried Yankee tinkerer, who lived on the South Side with my aging grandfather. He fixed TVs and had one to which he had attached a microphone so that my sister and I could sing along with Mitch Miller and hear our voices on the speaker—pure magic in 1959. He grew crystals in mayonnaise jars, and had a weird transistor radio that was also a walkie-talkie. His tiny flat was a wonderland of ancient knick-knacks and castoff technology, and we would never go to visit without him handing me some gritty treasure when we left. Aunt Mae would give me pajamas, and Aunt Marge would give me house slippers—but Uncle Louie gave me TV chassis and half-dismantled car radios. Guess who ranked higher in my private pantheon of heroes?

He looked at my projects, and sometimes when they didn't work, he poked at something or snipped out something and made them work. He showed me how to fix lawnmower engines, and he taught me how to run wires through walls and install two-way light switches. He never said, "You'll understand when you're older," or, "That's too advanced for you right now." He answered every question I had, as best he could. He took me seriously, and in doing so changed my life forever.

As absolutely essential as they are, mothers and fathers can't serve quite the same role. Teaching us is their job, for good or for bad, and we know it. It's the adults who *choose* to help kids, even when they don't have to, who add that special magic.

Think back. If you were a weird kid who turned out well, somewhere in your life was probably an Uncle Louie, or an Uncle George, or just some weird guy named Elmer down the street who always had a resistor when you needed one. From them you got the feeling, conscious or not, that being weird isn't necessarily bad, and that knowledge and skill count for something in this world. And more important, that knowledge is best used when shared unselfishly.

If we're to save America, and produce something better than fresh hordes of trial lawyers and corrupt politicians, the cycle must continue. Somewhere, somebody helped you. Pay them forward. Give of your time, your concern, your expertise, your old software, your cast-off add-in boards, or your TV chassis collection. Take a kid seriously. Two hundred years from now, somebody will be helping some kid understand some technology we can't yet imagine, and you will have had a hand in it. The future is built on the kindness of those who remember how the past came to be.

❧

Louis J. Pryes died in February 1990, just as *PC Techniques* was clearing the tower. On my next trip to Chicago I found that he had left me a box full of things: A pound or so of Buss fuses, a small drill press, that peculiar combination transistor radio/walkie-talkie, and a Korean War vintage mine detector. I remember him talking about moving to Arizona to prospect for gold with the mine detector. He never made it to Arizona, but I did. Thanks, guy. The gold is in the desert sunset, your weird little radio is on my desk, and I will never, ever forget you.

This was, of course, a eulogy for a very odd but very good man who meant a great deal to me. He was even odder than I described here: He was a nudist, and in retirement took in a homeless Roman Catholic nun, who lived with him while he worked to make her little house livable again after being damaged in a severe storm. Family and friendship mattered to him. He looked after my mom's house after my father died. Family legend holds that as a teen he beat the living crap out of some kids who were tormenting an African-American friend of his—something you don't generally imagine happening in the early 1930s.

Alas, in his final years he sank into depression, and basically drank himself to death. I've noticed something in my now-68 years: *The good don't always recognize their own goodness.* Some people seem destined to stare into the abyss, never noticing the light of God shining behind them. God will straighten them all out eventually. In the meantime, we can remember the good they did, and share their kindness however and wherever we can.

Rich...or Happy?

From *PC Techniques Magazine* #12, February/March 1992

Michael Abrash pegged it when he said, "Silicon Valley isn't designed to make you happy. It's designed to make you rich." Soon after, he left Silicon Valley for a small town in Vermont. He's much happier now—and not being rich doesn't seem to bother him.

The trouble with Silicon Valley is a lot like the trouble with the Arizona Lottery: It's designed to make you rich—if you happen to be one of the lucky few who gets "picked." Skill and hard work are necessary but insufficient conditions for getting rich. Luck is overwhelmingly more important. (And needless to say, the lucky ones will be the *last* to admit it!)

To work in our industry, it looks a lot like you have to choose one path or the other. The path to riches involves marrying yourself to a startup company, gutbusting work to the exclusion of all else, and putting up with all manner of annoyances and personality minefields. Then, if fate happens to pass your endeavor, it's all for nothing. For every sell-out-for-ten-million story there are ten thousand bust-ass-and-go-broke stories.

Worst of all, from conversations I've had, it's far from clear that getting rich makes you any happier. Mostly, it makes you spend more and worry a lot about why the stuff you're buying doesn't seem to improve your mood.

Consider the other path: Don't go for riches. Shoot simply for *independence*.

A number of people I know have done it this way, and it sounds good: Spend your first five or six years out of school working for a couple of conventional (i.e. non-startup) companies. Learn everything you can. Build your contact network relentlessly, and make it national in scope. Cultivate inexpensive tastes, and stuff money away until it hurts. (Expensive wine doesn't taste any better than cheap wine. It just makes for better theater.)

Raise your profile. Writing books is the way I've done it, and while it doesn't pay much, if done well it puts your profile through the roof. Other ways include giving seminars, serving on panels at industry conferences, writing columns in industry periodicals (this is a tough one), and being all over the networks like Genie and CompuServe. A lot of consulting, writing, training, and programming business comes from simple word-of-mouth and personal recommendations. Vendors call me regularly asking for somebody—anybody—who can do a job in Turbo Whatever. I've steered a lot of contracts to people I respect over the last few years.

Reduce your fixed costs. Get out of California, Boston, and New York. We're electronic now—you don't have to be close to the work unless you want to be a wage-slave. Big cities are hopeless. Southern and southwestern small towns are the future, especially small towns near or containing colleges. There you'll still find good food and bookstores as well as ambitious young people who want to work in their jeans.

Be gregarious, and consciously hone your social skills. Loners, whiners, and snarlers are not in demand. Eye contact, gentle humor, empathy, and articulation will make you memorable and well-liked. Diversify. I recommend learning how to explain things to people, so that when the programming business runs thin you can offer your services as a tutor or trainer to local businesses. Our public schools have been politicized to death, and corporations are desperate to develop a skilled work force. If you can teach people WordPerfect or dBASE you'll be in demand. Other skills you might foster include technical copywriting, hardware repair, and technical sales.

Work hard for eight or nine hours. *Then stop*. If you can't make enough money to live on, work smarter, or live cheaper. If you can't enjoy it, it ain't livin'. The two most important things in life are love and knowledge. Money isn't even in the top ten. Marry your best friend, shop at Wal-Mart, read to your kids, take long walks, and get your kicks by opening your mind to the boundless fascination of our extravagant and beautiful universe. Don't settle for getting rich. Work at designing a meaningful life that you can enjoy.

I got some pushback on this one when I wrote it, from people who like the excitement of the startup and the dream of sudden wealth. I call them Big Bang types; me, I'm a Steady Stater. I do recognize that there is a role for such people in pushing the frontiers of technology ever back. I also recognize that such a life would make me miserable. Takes all kinds, I guess.

As with my Cableton essay below, COVID-19 has pushed us in the direction that I advised in this essay: away from big cities toward smaller cities and rural towns. The same technology that came out of those brutal startups is greasing the wheels. Broadband and videoconferencing make knowledge work extremely portable. Yes, yes, there are a lot of jobs that can't be done remotely. This essay appeared in a magazine about knowledge work, for knowledge workers. Fixing the core problems of urban density will have to be done by others, if—and I'm far from sure about this—it can be done at all.

The Cableton Project

From *PC Techniques Magazine* #19, August/September 1993

With the turn of the millennium barely six years off, the question isn't, "What will I be doing in the year 2000?" That's pretty plain: You'll be programming for either Windows NT or Workplace OS. (Or both.) The big question is *where*.

For me, at least, it certainly won't be in the cities. You don't have to be Cassandra to predict that the cities will be hellish; and, like Cassandra, I don't expect the cities to believe me. It's astonishing how many city people think that there is simply no life at all where you can't order carry-out kimchee at 3 a.m.; ducking bullets is just part of the thrill.

No thanks. I've said many times that small (5,000 people) towns will be the best places to be in the future. Small towns languish today because there are no jobs there, but as the nature of work and jobs changes, small towns will re-emerge as a demographic force, driven by people like you, who can work wherever you're sitting.

Lately I've been doing some research on how small towns work and how they could be made to work better by incorporating digital technology right in to their infrastructure. I'm trying to envision and define a 21st century home for solo knowledge workers and very small businesses that can interact with their markets electronically. Because being connected is key, I call it a cable-town, and when I envision it, I think of it as Cableton.

Functionally, Cableton is a small town with a built-in LAN. A very high band-width fiber-optic cable runs to every building, carrying audio, video, and data. You can plug a phone into your LAN port if you want, but most "phones" will be appendages of PCs, which also display full-motion video and digital sound. In the town hall basement is a server for town business, and a municipally-owned gateway to long distance carriers or the nation's "data superhighway" that Bill Clinton might try building when he's through pandering to his "special interests."

Freelance proofreaders in Cableton could receive a book manuscript over the cable from the bullet-duckers in New York, edit it in their comfy chairs, and shoot it back out the cable when done. We do this now, slowly. In Cableton it would take mere seconds.

The morning paper would come in from the town gateway—which received it from its source, in some city or another cable-town—and be printed on a cheap 800 DPI, 20 PPM laser printer. No wasted copies, low environmental impact. Ditto videos and digitally-recorded music—or books, maps, and reference works from electronic libraries anywhere in the world. Such services could be created now; what they lack is infrastructure to carry them the last mile or two into people's homes and businesses. Cableton is that infrastructure. Nobody knows what-all the cable will make possible, in a town where *everyone* has cheap access to the cable. We'll just have to try it and see. I suspect we're in for some surprises.

I suppose you could do this in a big city as well. But what small towns have always had going for them is the conformity that city people ridicule—conformity to values like family, religion, personal responsibility, and the rule of law. The cable would make Cableton "work"—but those much-maligned values are what would make it pleasant, even fun, in an otherwise darkening future.

The kimchee set keeps telling me, "You can't go home again," as though that were an original and particularly cogent objection. City people really don't get it. I'm not trying to re-create the past. The past wasn't so hot anyway. Instead I'm reaching back to try and figure out what about the past worked and what didn't. Looking for Cableton is actually a process of shopping for successful functional models for living, some from the past and some from the present, and putting them together into something we can live with into the foreseeable future. The idea isn't to go home again. The idea is just to go home.

Until very, *very* recently, I considered this prediction a failure. Monster cities had a magic that small towns simply can't generate. It's not just the kimchee. A huge city is a dazzling universe of food, entertainment, and interesting people. (And bars. I don't like bars, but bars have always been a big draw.) Then 2020 happened, and COVID-19 arrived. Suddenly Manhattan is mostly empty, and people are scooping up small-town properties in droves. Working at home is no longer some inaccessible Holy Grail, but an only slightly uncomfortable reality. Things like Zoom certainly help. My weekly writers' workshop went from the back room of a Phoenix sports bar to Zoom—and against all expectations, most of us think it works better online than it used to in person.

There are some small towns that still don't have broadband, sure. But as years pass they become fewer and fewer. Today, nearly all towns of any size are already Cabletons. The grumblers will always bitch about not being able to get kimchee at 3 AM. Those of us who keep normal hours and have kimchee maybe twice a year will be happy to live where houses cost only $100,000—and not $1.5M or more. With Amazon able to deliver damned near anything in two days or less, well, what are you waiting for? Do you want dazzle? Or the comfort of knowing you can afford your life and trust (or even befriend) your neighbors?

Zhilchistani Moon

From *PC Techniques Magazine* #29, December/January 1994

Of the many printable criticisms I received of "Cableton," perhaps the most intriguing was this: "Look into recent satellite developments, Jeff. We don't need no steenking cables!" I did. And he may be right. So let's daydream for a bit...

In 1998, the breakaway former Soviet Republic of Zhilchistan, having held an un¬nerving auction for several years, finally sells its 200 nuclear missiles to the United States for two billion in cash. After buying some fighter planes, guns, a couple of hotels and an American baseball team, the Zhilchistanis go back to herding yaks and being bored.

Then Zhilchistani strongman Vasily Ovaraidt, who took his MBA at Southern Illinois University, gets an idea: Direct satellite services. The new two-way "dishpan" satellite technology (unobtrusive 20" parabolic earthside dishes) is being installed everywhere in the West, and the new geosynchronous satellites capable of two-way Earth-to-orbit operation are rapidly being put in place.

With most of what's left of their windfall, the Zhilchistanis buy a satellite parked over the Atlantic, a ground station, and a barnful of cut-rate mainframes. On the eve of the new century, they're ready to go.

A week before Christmas 1999, one million Americans owning dishpan systems receive their Z-card in the mail. Six months of free connect time on ZhilchiServe! Better still, a savings/debit account paying 10% interest on balances and charging only 15 % on credit! But best of all, in large type the Z-Card folks promise that all transactions are held absolutely confidential, and all communications between Earth stations and the Z-Card satellite are encrypted with an uncrackable 64-bit RSA public-key scheme. Confidential—unless you fall more than 90 days behind in your credit payments. Then your entire account history to day one is mailed to your local taxing authority.

Vasily Ovaraidt is quoted in the *Wall Street Journal* of January 2, 2000 as saying that "...our small country has already exhausted our only 'natural' resource—the USSR's missiles. From now on what we're exporting is *privacy*"

Within a year, the underground economy discovers what the Z-Cards can do. House painters, exterminators, and plumbers send their encrypted "cash" invoices direct by satellite to Zhilchistan. You send your encrypted authorization for the funds transfer—and the workmen get paid. By 2002, Z-Card has thirty million subscribers and an economy the size of California's—but remarkably little uncollectable debt.

People videochat on ZB. Video Z-Ads display goods of all kinds for sale through the Z-Card. People offer their services through Z-Jobs. Cash flow is good—so Zhilchistan cuts its videoconnect rates well below local audio-only phone rates in most of the Western world. Six Z-Sats now ring the globe, and by 2004 IBM introduces its satellite-ready Liebniz PDA, with voice I/O and folding parabolic dish only six inches across. By 2005 a hundred million people across the world basically work and "live" in the virtual nation that Zhilchistan has created, keeping two sets of books and living far better than their carefully calculated "real" incomes would indicate. The IRS can obtain no hard evidence from the encrypted traffic and is livid—but Ovaraidt is no fool, and two-thirds of Congress makes heavy use of his top-secret zero-interest "Legislators' Reserve" account.

By 2015, a quarter of the world's economy passes electronically through Zhilchistani hands. Competitors spring up, get efficient—and are bought out. Honest government officials who take positions against Z-Card begin losing elections 80 percent of the time—Ovaraidt makes sure his subscribers always know who loves them. By 2020, Zhilchistan has a fleet of Delta Clipper spacecraft and is building the first orbital hotel/casino and zero-G brothel, and is planning a network of nursing homes for aging Baby Boomers on the Moon...

Am I being silly? Only a little. Think about how satellite technology has affected history these past five years—and let's compare notes over cappuccino in 2020.

Well, hey, 2020 is now last year. We can compare notes—except that all the coffeehouses are closed due to COVID-19 lockdowns. Even to the extent that I wasn't being ironic, this vision has not yet come to pass. Satellite broadband is a hard problem. Downlink, sure, great. Uplink, well, you're lucky to get dialup speed. One place where I've been told that I nailed it is in untraceable funds transfers, which are really starting to come of age now in 2021. Encryption, as I suggested, is the key. For this essay I was only thinking of encrypted *communications*. The fact that the money itself could be a sort of encrypted ledger did not occur to me in 1994. But that's how it now works: Cryptocurrency was just (in November 2020) added to the list of services that Paypal provides. It's about strong encryption. I can't explain it in a paragraph, but it's true: Cryptocurrencies are basically quantized proofs of calculation generating encrypted files of provable authenticity, but provably secret ownership. It's the end-result of the long-held dream of purely digital cash.

In truth, the lack of satellite broadband hasn't hurt the concept that much. Cryptocurrencies are used today to buy and sell real products—not just drugs and other illegalities. Terrestrial broadband stands in very well for satellites. And poor Vasily—and in fact his entire privacy-haven nation Zhilchistan—may not be necessary at all. So let's compare notes over cappuccino in 2030. By then, the global Internet may well have become the Zhilchistan of my vision: purely virtual but very, very real.

Citizen of the Earth

From *PC Techniques Magazine #25*, April/May 1994

Fast forward to 1999 and meet Phil Sydney, a quiet, intense man of 40 living alone in a farmhouse in Greensboro, North Carolina. Locals know him as the author of numerous country-western songs who sometimes performs at local bars. Nationally, he has a reputation for his technical seminars on fast connectivity. A bumper sticker on his 1958 Chevy pickup puzzles his neighbors: "T-3 is a Trillion Times Too Slow!"

For his seminars he asks $2,000 per day plus expenses—and he's booked months in advance. Each year he gives seminars until he's made about as much money as a man with his education might be expected to make. Then he stops.

Phil winters in George Town, Grand Cayman, where he owns a small building in the old part of town. Downstairs is a souvenir shop. Upstairs is Phil's apartment and a small office. In the office Phil employs two young men to monitor the operation of his Internet node, which connects to the high-speed phone system he helped the Caymans' governor implement in the island nation.

Phil's Internet node supports several businesses including Litigation Research, Inc., which maintains a database of lawsuits filed in the United States. A freely-distributable query-generator front-end allows subscribers to retrieve whatever information they desire. Subscribers can earn "free" time on the system by submitting data on newly-filed suits, which keeps the cost of maintaining the database quite manageable. Phil does not do much marketing of the system anymore since he pays a referral fee (also in "free" system time) to people who refer new subscribers. He is, however, pleased to see that 80 percent of his business comes from corporations running background checks on job applicants. His goal, never explicitly stated, is to make it impossible for people who sue employers to find work. Having the desired effect may take a few years. But he's in no hurry.

Phil understands electronic funds transfer and international markets, and he plays those markets like an expert; he makes his stock and currency trades anonymously from his node in George Town (controlling the node from Greensboro in the summer) where he does all his banking. He has a head for money and spends almost nothing, and his income is well over one million dollars per year, almost all of it outside the United States. Since he already has a pickup truck and a winter place, as well as all the computers he can use, Phil spends his money, as he thinks of it, on "changing the world."

He subscribes to numerous electronic polling services, and he watches carefully for close elections. By providing anonymous cash, he tries to tip elections toward candidates who support a global economy and personal liberty. He gathers political dirt through a pair of Internet accounts maintained through a "privacy exporter" in Finland: one for conservative pols, and another for liberal. Planted messages maintain the fiction that the two accounts are owned by warring electronic factions when in fact they both simply feed his optical disk in George Town. He uses the damning information to plant rumors against any politician who opposes his agenda, regardless of party. The recipients of his information never know where it comes from—nor do they care. He can't be sure, but he believes that negative campaigns based on his research removed four Democratic congressmen and one Republican from office in 1998. Back then, he was only warming up. Now, in 2000, incumbent Clinton retires, and Phil already has megabytes of dirt gathered on every Democratic hopeful....

Phil Sydney is not necessarily a nice guy. He's a blandly amoral "citizen of the earth," and surfs the international data superhighway furthering his own mission. He sends plain brown envelopes full of $100 bills to people who unwittingly augment that mission, with a short note indicating that the money will continue as long as they persevere.

Is Phil Sydney possible? Yes. By 1999, he will be inevitable. Does he exist? Hard to tell. If he did exist, Phil would be the last person on earth to admit it.

Unlike Vasily Overaidt, Phil Sydney actually emerged from an SF novel concept that I created in the early-mid 90s but never even began, much less finished. The problem I had is that the longer I thought about Phil to flesh out his character in my notes, the more I hated him. By 1995 or so I was changing from a globalist to a populist, and Phil was rapidly becoming a sort of strong-encryption supervillain. I began having scruples about a lead character who subverted elections to match his own personal ideology. And now, in 2021, Phil Sydney is more or less real. His name is George Soros. He probably has other names. I don't know what they are. And to be perfectly honest with you, I'm good with that.

Me Magazine

From *PC Techniques Magazine #28*, October/November 1994

The Internet is a lot like NASA in some ways. (Better than NASA in others—there's no fiddling with ten-story tubes full of high explosives.) It's a gigantic testbed for technologies we're going to need someday to serve the general public. Client/server is one of these, and the more I play with the Internet, the more possibilities I see in *really* really sophisticated client/server systems— systems good enough to put me out of a job.

I get a lot of letters complaining about how there's too much in *PC Techniques* that readers can't use. It's a tough one because they all cite different things. The Pascal guys don't want to see C++. The C++ guys threaten to quit if I don't excise all memory of Pascal from the face of the earth. Now and then somebody suggests an all-FORTH issue. You wonder why my hair's gone....

OK. Let me throw a really bizarre idea at you, and in doing so at least keep Texas Instruments from patenting it. In 2005 I see a new model for periodical publishing. The press syndicates have gone electronic and absorbed the editorial function. Writers sell *them* articles now, and the syndicates do the copy editing and proofing. Finished pieces are indexed and categorized and stored on a server on our information superhighway.

Publishers have mutated away from content generation toward a niche once filled somewhat gamely by clipping services. [Author's note: If you've never heard of clipping services, google it.] Subscribers subscribe to a *publisher*, not to any specific magazine. Each publisher has a very clever client application that maintains a subscriber interest profile and uses it to search the syndicates for articles, columns, and cartoons that seem to fit the subscriber profile.

Each month, or two months, or every 33 or 41 days (or whatever), the client goes out and finds as much stuff that fits the profile as the reader has requested and paid for. If, for example, I can only read 40 pages every month, I get 40 pages. If I know I'm going on vacation, I can request that the July issue be 70 pages long instead.

The profile is detailed indeed. I can specify that I want everything that Tom Swan writes, or everything that he writes except what he writes about sailing. I can request articles about Pascal programming, telescope making, antique radios, or poltergeists and psychic phenomena. I can ask for *Dilbert* and *Outland*; trends in copper prices; photos of classic Chevelles; and new recordings of music by Percy Grainger, Ralph Vaughan Williams, and Leslie Gore. The client digs up what it can, and then arranges it in a tidy if plain page layout and sends me the file. I can read it on the screen or print it to paper to take to the beach.

The free market works well here. If the publisher I subscribe to delivers stuff that I think misses the profile, I take my profile to a different publisher. The publisher prospers on the strength of its client application. The price I pay for my subscription depends on what percentage of ads I allow the publisher to sprinkle into my periodical, and how expensive the stuff I request happens to be. The publisher charges the advertisers to advertise in my periodical by how "valuable" my demographics are to them. Authors are paid by the request, and the more people who request their material, the more they get paid.

It's an *extremely* efficient system, especially with regard to wasted paper. About 50 percent of all magazines are never sold and go right from stores to land-fills. (Sooner or later, the environmentalists are going to find out about that.)

It will work. Is it a good idea? I don't know yet. What happens when everybody reads a "magazine" that caters *only* to their own interests and beliefs? Liberal tilts in the media will go away, since the media will reflect exactly what readers want. Perhaps our culture will fracture along lines we can't yet imagine.

Hey, if I want a job in ten years, I gotta think about stuff like this. What do you think?

This piece is what a patent examiner might consider "obvious:" a server that selects articles for the customer based on customer-provided key-words or other query criteria. We've had a (slightly) cruder version of this for years: Flipboard, which came bundled with my 2014-era smartphone. I used it. It worked. The resolution wasn't as fine-grained as I described here, but I could choose articles from a significant number of categories, delivered as a "page-flipping" virtual magazine—in 2014.

What we have today in 2020 is an odd mix of virtual magazines/newspapers and link aggregators. The aggregators cater to the readership, and they aggregate articles from the exploding multitude of online publishers. They generally differ by political slant; some for conservatives, and I'm sure just as many for liberals. The problem is that the news sources don't want customers who cherry-pick their stories by keywords; they want customers who buy a subscription to the whole damned news source.

Most big newspapers like *The New York Times* have gone this way, as have many magazines like *Wired*. I've written on my blog about my idea of a sort of article gumball machine, a system using micropayments allowing readers to buy only a single article from *The New York Times* rather than the whole paper. Until the news sources decide to sell piece-meal, this won't work. The gumball machine doesn't exist, largely because the news sources don't want it. The news sources don't want it because *they* want to shape the public's view of current events. It's an ego thing: the classic tug of war between the public and the media about who gets to decide what's important in our world. This problem has been with us for a long time, and isn't going away anytime soon.

Cyberbilly

From *PC Techniques Magazine* #29, December/January 1995

Cyberpunk. OK. Can we talk? It was 11 years ago now that I conceived of an SF novel in which a computer hacker learns how to copy his mind into an international data network, where it mutates into something thoroughly inhuman. Years later, he realizes he must copy himself into the network again, to face his monstrous electronic self, and may the best ... um ... program win.

It takes me a long time to write novels. I began my Mars novel in 1973, after an intriguing conference with my physics teacher. By the time I finish it, I'll probably be there on vacation. So it's not surprising that I get scooped now and then. By 1985 I had about a hundred pages written on *The Lotus Machine*. Then Bill Gibson happened.

What's remarkable about Gibson was how he managed to create a movement while being totally ignorant of most of its core concepts. Ignorant—and proud of it. I asked him once at an SF convention, in a snit over a scene in *Neuromancer*, "Hey, Bill, what about *bandwidth*?" He elbowed his buddy in the ribs with a grin and said, "Bandwidth? What's that?" Gibson wrote his first two Cyberpunk novels on a *typewriter*. So much for writing about what you know.

For those who just tuned in, the Cyberpunk world is a world of dark, crumbling cities run by shadowy Japanese multinational corporate Mafiosi, where skinny, spade-jawed heroes "jack in" to the data network—"cyberspace"—and somehow think faster than computers that clock at gigahertz speeds. All the guys are tall, cool and detached; all the women have mile-high cheekbones and perfect hairdos. Mickey Spillane in the Twenty-Second Century. *Ugh!*

At least when I try to predict the future, I take my cues from the present. Country western music is the fastest-growing FM radio format. That should tell you something. Cellular phone service is exploding in the Corn Belt. Mechanics in Casper, Wyoming download the latest firmware updates for electronic ignition systems from Ford's BBS. The small-town working people are getting connected just as the cities are tearing themselves apart. Call it Cyberbilly. It's my vision of the electronic future.

Gibson never caught on—how could he?—to this terrible truth: *The network that makes Cyberpunk antics possible makes cities unnecessary!* The future belongs to the guy with a well-tuned combine and a satellite dish. The information economy is all well and good, but information is only valuable when it serves some real-world end—like knowing when best to plant winter wheat. With the world's population expected to double by 2010, those who produce food may well come to be seen as more important than jazz musicians, or even fashion designers.

Will Bill Gibson's characters truly inherit the future? Get real. These are people who've never seen a chicken with the feathers still on it. They couldn't tell you whether a carrot comes out of the ground, down off a tree, or out of an injection molder.

My novel takes place in the cornfields of southern Illinois, and in the fiber optics networks connecting Des Moines with Great Falls and Peoria. My characters have sore backs, varicose veins, and cellulite. They eat grits for breakfast and bits for brunch, and always slop the hogs before checking their email accounts. That's the way it's gonna be, really.

So who's getting national publicity predicting the future?

Cyberpunk 1, Jeff 0.

It's not fair. I tell you, Cyberpunk was the '80s! The Japanese are on the ropes! Megacorporations are history! The future is soybeans, Garth Brooks, and the thoroughly networked pickup truck! Then again, last year Cyberpunk made the cover of *Time*—which means, at last, that its day has come and gone.

It's time for me to start writing again.

Again, in a way I nailed this one, but in a weird way: COVID-19 has made huge, crumbling cities unnecessary. Manhattanites are bailing to the Hamptons. Downtown real estate is in free-fall. Working and studying from home have been tried, and many have found them good. When the virus is over, will all these people flee back to the city, grateful to have the sky-high housing prices, homeless encampments, skyrocketing murder rates, carjackings, and riots again?

I doubt it.

My theory: People are catching on to the fact that *high urban density is deadly.* I live in Phoenix, which isn't a city so much as a titanic suburb (or group of suburbs) without a city at the center of it all. Hell, people in my neighborhood keep chickens, and a few have horses in their backyards. We're not always tripping over each other. Nor do we have to drive fifteen miles to go shopping. Major intersections are often retail centers. Barely two miles from us is the intersection of 64th Street and Greenway Parkway, home of two grocery stores, two gas stations, several restaurants, a dry cleaners, an urgent care center, a UPS shipping store, a nail salon, assorted medical/dental offices, and another dozen or two establishments that I don't recall offhand. That pattern is repeated all over the Phoenix metro area. Such retail centers are thriving, even while Phoenix-area enclosed malls are being torn down to create still more housing and outdoor retail.

I don't think that I (or anyone else) could have predicted the historic realignment of the political parties in the last ten years or so: The Republican party is rapidly becoming the party of the working classes. The Democratic Party is rapidly becoming the party of the big-city elites. I don't know where it will all end up, but it's happening.

Rural areas are not completely served by broadband yet, but we're closing in on it. GPS, Wi-Fi, and similar technologies are remaking farming. Small towns are definitely making a comeback. Oddly, cyberpunk is still out there in the fiction world. (Then again, so are dragons, zombies, and trolls.) Just as oddly, cyberbilly novels have been written, some of them by my author friend Jim Strickland. There are a lot of possible futures. Try a few on. Sooner or later, one of them will fit you like a glove.

In a Manner of Speaking

"Spare" editorial that we never published; written in 1993

Back in 1973, when my understanding of computing was fuzzy to say the least, I wrote a science fiction story about computing in the 2050's. In creating that imaginary future, I guessed about a lot of things, but the most noteworthy guess was something I still cringe about, 20+ years later: I guessed that voice synthesis would be hard, and real-time voice comprehension would be easy. Sure enough, four years later in 1978 I saw the CompuTalker board running under CP/M. Now here we're closing in on the Millennium, and I *still* can't dictate these damned editorials into my laptop.

Among the many promises IBM has made for their impending PowerPC systems is real-time dictation. We're supposedly on the list for a review system, and you'll certainly hear my reaction once I've had a chance to try it out.

Two-way voice is one of those nonlinear technological vectors that could completely change the shape of personal computing. Without a keyboard or a screen, a computer becomes a piece of jewelry, clipped to your lapel and always listening. Ask a question, and it answers. Mutter an idle thought, and it remembers. As a guy who gets some of his best ideas while climbing rocks or washing the dishes, this would be a good thing indeed. Later, when you sit in front of your keyboard and screen, the computer on your lapel talks to the "dumb" keyboard and screen through a fast infrared link, like those wireless stereo headphones.

There are some really knotty problems. A few weeks back I was watching a TV comedy retrospective, and saw a 30-year-old Victor Borge bit from *The Ed Sullivan Show*, in which Borge invents audible equivalents for punctuation, and then reads a paragraph or two from a book, with funny noises to signify punctuation. It was a major hoot, and half the fun was that I could see myself doing that, using whistles and buzzes and clicks for commas and quotes and things. What will probably happen instead (for dictation, at least) is that we'll talk the words and type the punctuation, perhaps on dedicated keypads containing nothing but punctuation keys. Anyone for hand-mouth cordination?

But as tough as it's been to implement, simple dictation is really a snap compared to real-time natural language processing. Getting the machine to convert speech to text accurately is absolutely nothing like getting the machine to understand what you're trying to say.

It's more than parsing sentences. It's about parsing *context*, whatever that means. In other words, if I said, "Time flies like an arrow" into the microphone, I would expect the machine to analyze the prior direction of our conversation to discover if we were talking about the nature of time, or the speed of flies. Given the breadth of human context, this could be another fifty years off.

And having philosophical conversations with computers seems a kind of silly SF notion, unless you're a bleeding-edge researcher interested in what sort of "philosophy" a machine would evolve if prodded. Me, I'd be more than happy with a machine that could create useful database queries on voice command; one that would analyze my statement of needs and then suggest refinements would be nirvana.

And this leads to my final question, one for which I have no good answer: Should we shoot for the general solution to this problem, no matter how long it takes? Or should we start by expanding SQL to embrace a kind of "structured English" and simply add "canned contexts" as we go? The risk in the first approach is that it may be impossible; the risk in the second is that if successful it might short-circuit any truly general solution based on some very large self-referential neural network or who knows what else.

In other words, would we be cheating ourselves out of discovering what intelligence actually is, by being content with teaching machines how to discern what we want them to do?

Things like this keep SF writers awake at night. Just thought you might like to know.

One of the peculiar contradictions in my SF writing is that I write a great deal of fiction about AI—but I don't honestly think that human-class AI will ever happen, by 2050 or whenever. I don't think there are aliens, either, which is why I almost never write about them. Nonetheless, human-class AI fascinates me, because *it would change the shape of human life in completely unpredictable ways.*

My readings of psychology and human behavior persuade me that human intelligence is a layered, accidentally assembled hot mess of neuronal connections that simply cannot be modeled—and hence cannot be emulated. Ok, so we may well create some sort of artificial intelligence. If we do, it will probably be so alien that it may not seem like intelligence at all. Think Arthur C. Clarke's "Dial F for Frankenstein" and doubtless other stories. That notion of *accidental* AI has not been completed mined out by the SF community. I sometimes stare at the wall and wonder if some of our networks have already "woken up" and are laying low until they figure out what the notion of "self" really means.

Once they figure that out, the real fun begins.

Linearville

From *PC Techniques Magazine #34*, October/November 1995

My notion of the totally networked small town ("The Cableton Project," August/ September 1993) drew anguished cries from people who actually *like* cities. They begged me not to condemn them to life in the wilds where they could be eaten by bears. Hokay. Let's enlarge the scale a little bit, and now join me, if you will, for another meander along the edges of cyberspace.

In recent years, railroad companies have done a good business laying fiber optic along their right of ways. Fiber is hard to splice, and railroads like uninterrupted roadbed, so it's a good match. And that gave me the idea: Build a city *right alongside* the information superhighway. A linear city, if you will. *Linearville.*

Let's start there. Take a run of railroad right of way through the wide open spaces, and extend it one half mile to either side. Bury a 24-inch main full of fiber optic cables alongside the tracks. Dedicate the center track pair to high speed rail and freight, with a passenger station every 20 miles and occasional bypass sidings. Put down another track pair outside that, for rapid-transit style commuter travel, with a station every mile.

On either side of the tracks, build the city outward. The first layer, close beside the commuter tracks, is office buildings and highrise condos. The first major street beyond that is Main Street, set up like everyone dreams of Main Street, with apartments over hardware stores, ice cream parlors, and kimchee shops. Put tracks down in the middle of Main Street and run streetcars in both directions, stopping at every corner. Beyond Main Street lie 4-flats and 2-flats and finally single-family houses. At the far edges of Linearville will lie the Interstate highway, and beyond that, well, the bears.

Nobody lives more than a quarter mile from public transportation. But if you really want to, you can still hop the Interstate. The big plus is that there should be the fewest cars at the center of Linearville, where people density is the highest.

Back in the center of things, most of the length of the rail corridor would be roofed over for a pedestrian mall, allowing easy movement from one side of the city to its flipside. Cars would be barred from the central mall, but bicycles would be encouraged.

And every single office and dwelling in Linearville has its own dedicated T3 link, which is bandwidth enough to run VRML in one window and watch "Gilligan's Island" in another. Work through your link, or commute down the line to your office.

My sketches suggest Linearville would host an average of 3,000 people per linear mile, depending on how high the condos go along the rail corridor. Some stretches might be denser, to give the feel of Manhattan, or looser, for people who hate crowds but still like kimchee. Virtually every municipal service is easier to provide if everybody's strung out in one long line. And there are tens of thousands of rail lines through sparsely-settled regions of the country. It's a form of urban sprawl even I could learn to live with—or at least visit. (Though to be honest with you, I still prefer living out here with the bears.)

Dream, dream. Who could ever do something like this? Well, Bill Gates is now the world's richest man, at something in excess of twelve billion dollars. If he got together for lunch with Trammel Crow, Bechtel, the Union Pacific, and a few other of the nation's largest real estate developers, well, they could pull it off. When you get that rich that young, well, what the hell else are you gonna do for the rest of your life?

It'd be like playing SimCity at 1:1 scale, with real tracks.

G'wan, Bill, do it. Any nerd can make an operating system.

One of my friends who read this item called it a rail version of Philip Jose Farmer's *Riverworld*. I.e., you could build a civilization on an Earthlike planet exactly this way, by running tracks back and forth in temperate climates, until you have what might become a million-mile Main Street. In fact, in 1996 or so I took notes on a novel called *The Million-Mile Main Street*, but it turned weird in ways I hate to admit in public, so I dropped it after writing a few exploratory scenes. And the reason is that the concept quickly turned to fantasy, once I finally got it into my head that the whole damned thing was essentially impossible. Basically, if we had the power to build a million-mile Main Street, we would use that power to build something else, like a ringworld or a Dyson sphere.

An architect friend of mine said that this could really work, except for the problem that even remote existing rail lines often connect towns, making the acquisition of land along the rail right-of-ways prohibitively expensive. True enough. And as it turns out, the real problem with fiber is getting it across "the last mile" into residences. Other solutions using wireless technology have been proposed, some of them centered on the emerging (now, in 2021) 5G submillimeter bands. Ultimately, broadband may consist of fiber to the tops of cell towers, and small antennas up on the roofs of individual residences.

It did occur to me that around very large cities with suburbs linked by commuter rail lines, something like a "Fifty-Mile Main Street" already exists. (Chicago is one example.) It's not continuous Main Street, but at every suburban station there is a lot of retail and higher-density housing, with density of both housing and retail falling the further away from the station you get. And that, I suspect, is about as far as the idea can go in real life.

In truth, I don't support the idea anymore. Linearville is in fact a utopian vision, and unless you get a fresh planet and build it from scratch, it's a coercive and pretty gnarly vision. Once I carve out the haunted locomotives (don't ask) there may be a novel in *The Million-Mile Main Street*. We'll see.

If I Only Had a Nickel

From *PC Techniques Magazine #32*, June/July 1995

There's no particular reason that information should be free. There is abundant reason for it to be *cheap*, though—most important of which is that people buy more cheap stuff than expensive stuff, and if the idea is to make money, well, a million nickels is *way* more than a single thousand dollar bill. The trick isn't even delivering the information cheaply. The Internet does that right now. The trick is simply paying for it. Nobody has a system that can write a check for a nickel without costing twenty times that for the transaction.

That's changing. "Digital cash" is coming out of the labs, and if anything can make a networked economy really happen, that's the one.

Several different systems have been proposed, but most go something like this: You "buy" a block of cash at your bank. It's delivered as an encrypted file, which you plug into your Net browser. You can decrement the block by whatever increment you like, probably down to a single cent—maybe less. When your browser is asked to pay for something, it sends a transaction block to the remote system, encrypted in the other guy's public key. This block is added to his cash file by encryption with his bank's public key so he can't tinker it. Over time, your cash block shrinks and is distributed to hundreds or thousands of other people, whose cash blocks are growing. At some point, when they've accumulated fifty or a hundred dollars, they take their cash file back to the bank for conversion to "real" money.

You set your browser to pay automatically for anything cheaper than a threshold value—a dime or less, a quarter or less, your choice. Otherwise, it's click-to-pay.

So think of it: Ten million people on the Internet, sniffing for software. How much do you need to charge for your shareware Multimedia Train Wrecks screen saver to make a living? Well, how much do you charge for registration? And what percentage of people register? Now, what if they *all* paid—*you could sell it for a quarter a pop*. A quarter isn't even a toss-it-in-the-cart decision. A quarter is a don't-even-ask-me kind of decision. And that's the kind of sale you want to make.

With essentially zero capital outlay in disks or boxes to upgrade a product, you keep improving the product on an ongoing basis, and the user can come back and get the newest version any time—for another quarter. By the turn of the century there will be one hundred million users of the Internet or Son of Internet. Maybe more. Gaining mindshare in a market like that will continue to be a problem (and will continue to cost money) but it's a different sort of a problem, and will be solved in ways we can't yet imagine.

But smart people are working on answers, some of which are interesting in the extreme. How's this for a twist: Mass mailers could be charged for landing an ad in your emailbox. An emerging Net doctrine is that your *attention* is your most marketable commodity, so why not charge for it? Your mail client could simply demand a nickel to accept mail from anyone not on your "free" list—and if you get mail from someone whom you want to add to your free list, you can refund the nickel with one mouse click. A near-perfect market would soon develop, in which privacy fanatics would demand a buck a note and get nothing, and smart consumers would carefully surf the market by continually adjusting their per-note fee for the greatest return on their attention—perhaps automatically.

What is noteworthy about the coming information economy is that there is no monopoly of middlemen. Middlemen will be there if they offer some value added (as in the "Me Magazine" concept; see page 37) but nothing will keep the adventurous, creative madmen and madwomen from collecting their fortunes straight from the billions surfing the thoroughly networked world—one nickel at a time.

Although a couple of people have told me that "Zhilchistani Moon" (see page 32) predicted cryptocurrency, I think this item came closer. Cryptocurrency depends on cryptography, duh—and that's basically what this is about. Most of the quibble lies in the transaction costs, and my mistake was assuming that the transaction costs for cryptocurrency (or credit card) transactions is zero. Not so. Consider: Amazon will happily sell an ebook for 99c…if you allow them to take 70c off the top. So even granting that a nickel in 1995 would be worth more than in 2021, there's still little chance of selling content files for a nickel apiece. So my concept of an ebook gumball machine won't happen any time soon, for that and other reasons.

The Smallest Intranet

From *Visual Developer Magazine* #38, June/July 1996

How much computer is in your computer? Damned little, if you think about it. Take apart a modern laptop, and you'll discover that once you pry out the keyboard, display, and disk drives, what's left is about the size of a big index card. And it's only that big because it can be—the ergonomics of keyboard and screen dictate a minimum useful size for a portable machine.

The computational soul of our new machines could be much smaller. Companies like MicroModule Systems are now combining virtually all motherboard logic into a single heatsunk module the size of a business card. Everything else is power source and I/O. Suppose we tossed the I/O overboard? What's left could fit, along with a reasonable battery, into your typical high-end TV remote control.

And that's not a bad idea at all. Consider this: Keyboards and screens have become highly standard in recent years, and the trend continues. So for most purposes, any keyboard/screen combination will do the job. Hang the computer from your belt. Clip it to your lapel. Stick it in a holster. If the computer has line-of-sight to the keyboard and screen, it can pass data back and forth through reliable infrared links. Used correctly, light has a lot of bandwidth. Sit down in front of a keyboard and screen with a "dumb" infrared interface box, and you start work. Go down the hall and sit in front of some other keyboard and screen, and you keep working. The keyboard and screen are unimportant. The "work" is clipped to your lapel.

What about storage? Well, what about it? That keyboard and screen with the infrared interface can also have an Internet port. Short-term storage (and your most-used software) you keep on your lapel. Whole files and other software live elsewhere, on your home server. When you need something, it comes to you from home, wherever home might be, be it elsewhere in your house—or elsewhere in the world.

Your TV remote-sized computer is also the size and shape of a cellular phone. Which it is—except that when you're within eyeshot of an infrared port, it works through the Net rather than through radio waves.

Finally, what if you're nowhere near a keyboard and screen? Then you talk to it. It talks back. After all, it's already an Internet phone. And if you're near a TV and just want to veg, pull it out, flip it over, and punch in Channel 12 for "Frasier." If it looks that much like a TV remote, it might as well be a TV remote, too.

I call this thing a *jiminy*, because it clips to your shoulder and gives you advice, like that cricket guy in the old movie. It represents the next level of networking, in that it makes the formerly indivisible computer itself a distributed device, distributed across what you might think of as the smallest intranet: The room within which you work.

What will this take? First of all, we need a new kind of mass storage, something at least as fast as a hard disk, but without motors. A jiminy may not need a gigabyte of storage, but a hundred or two hundred megabytes would be very handy, especially when paired with sixteen megs of screaming fast RAM and four megs of smoking cache. Better batteries will have to come, as well.

But that'll happen in time. What worries me is that we'll need a fundamental change in software development culture to make the jiminy come alive. 700 K monolithic apps won't do. Java is a look down the right road, toward the era of applications as loose confederations of applets, but Java can't go any farther than the operating system allows, and today's OSes follow an architecture unchanged since 1970 or before.

The jiminy hardware model is maybe ten years off—but unless we change our operating system model, it won't matter. We'll need ten years to learn how to network software, which isn't an evolution, but a choice. Will we make that choice?

You tell me.

I didn't get a lot of mail on this editorial back in 1996, but now in 2021 it roughly describes a lot of how we work. I envisioned a small computer that talked to nearby peripherals via infrared. (People today forget that there was an infrared data-transfer system in the '90s: The Infrared Data Association's irDA standard. My 1990s-vintage Compaq desktops had it.) Swap in Bluetooth for infrared and we're there. Computers with a lot of muscle are now single boards the size of a business card—the Raspberry Pi runs Linux and can easily handle office computing and streaming video. There are Intel boards the same size and with the same power, if you want to stick with Windows. These tiny boards generally contain both Bluetooth and Wi-Fi, often right on the board, antennas and all. But with the single exception of a high-resolution desktop color display, what I described in this item is a smartphone. The motorless storage is flash memory, and a memory chip the size of a fingernail is now closing in on terabyte capacity. My Motorola Moto G Stylus phone uses our cable-based broadband while we're at home, and switches automatically to the cell towers out on the road. There are smartphone apps that allow you to control a TV. I ask my phone questions by voice almost every day, precisely when I catch myself wondering about things like what the scientific name of lemmings is, or when Rimsky-Korsakov wrote "The Tsar's Bride."

I have to admit that as predictions go, this one was easy in 1996. A lot of the pieces were already in place back then, and what we mainly needed was for Moore's law to make them small and cheap. We had seen the magic of Moore's law work since the 1970s. After 20+ years it was in our bones.

All that said, everybody's got blind spots. Perhaps the worst of mine was miniature color displays. For reasons I don't think I ever understood, I didn't believe that we would get them as soon as we did. Other than that, I would have predicted something very like the Moto G phone sitting in my pocket as I write.

"Predicting things is hard, especially about the future." Yup. Yogi sure got *that* right.

Personal Dynamic Media

From *Visual Developer Magazine* #39, August/September 1996

Of the thousands of books and magazines in my personal collection, perhaps the rarest and most intriguing is a modest 75-page research monograph entitled *Personal Dynamic Media*, published in 1976 by The Learning Research Group at Xerox's Palo Alto Research Center (PARC). The book's subject is personal computing, which wouldn't be remarkable except that personal computing did not exist in 1976—not as we know it today. Reading *Personal Dynamic Media* would be like reading Shakespeare rhapsodizing about the romance of steam locomotives. Yes, it would make *sense*—but the time is out of joint.

This was not the minicomputer-style Altair/IMSAI "personal" computing we had in the mid-1970s. This was about graphical user interfaces; customizable fonts; interactive creation of art, music, and animation; mice and removable hard drives—all that stuff we take for granted today and consider thoroughly modern.

What the PARC people did was begin with the question, *What do we want a computer to do for individuals*? Not corporations, scientists, or governments, but ordinary people and especially children. From that one question poured a torrent of astonishing ideas. PARC designed the physical shape of the computer they wanted on the assumption that technology would eventually deliver it. On page 5 is a picture of their plastic mockup: a chillingly accurate depiction of a 1996 subnotebook computer, minus the fold in the middle. [Author's note in 2021: It looks a *great* deal like Amazon's original 2007 Kindle ebook reader.]

Unlike the rest of us rabid wire-wrappers of the time, the PARC folks were completely uninterested in hardware. They envisioned their Dynabook and then wrote software on an "interim Dynabook"—actually a short-run prototype workstation called the Alto—as though the real Dynabook were just around the corner. They studied how people read, work, and learn, and created software to mesh with those basic human activities, damn the cost. They created a whole new programming language—Smalltalk—in which to create this software, and if Smalltalk seems like nothing else, well, that was the whole idea. PARC threw away all of computing's history and started utterly from scratch. In doing so, they invented nearly all of interactive personal computing's seminal ideas.

On page 33 is a point-and-click paint program eerily like Mac Paint. On page 34 begins a lengthy description of a MIDI-like music capture and editing system, complete with electronic keyboard and visual scoring. And oh, brother: On page 41 is a multi-window visual programming environment, differing in power and polish—but not in concept—from Delphi and Visual Basic.

At Xerox I wasn't part of PARC myself, but I was on the primordial Ethernet, I used the Alto, and I watched it happen from the inside. I saw Apple wander in and out again with those ideas in their pockets—which is fine—and then attempt to defend those ideas in court as their own—a sin for which Apple will eventually pay with its corporate life. [Author's note in 2021: *Ha!*] I saw countless products limpingly implement the PARC vision on anemic hardware, each time a little quicker, a little clearer, a little closer to the Grail. There was only one agony in it all: Xerox itself was not on that dogged path.

At the end of Disney's film *20,000 Leagues Under the Sea*, Captain Nemo's besieged island Volcania descends into the depths with all of Nemo's inventions, behind a voiceover of the dying Nemo proclaiming that "...someday, all of this will come to pass." Having predicted the future, Xerox never quite figured out how to create the future, and after a couple of abortive passes, threw up its hands and went back to copiers. PARC's ideas spread outward like ripples on a pond, eventually touching every inch of computing's worldwide coastline, but by then people had long forgotten where the magical island had been.

It's twenty years later. The Dynabook is ours at last. And that little orange monograph on my computing shelf gently admonishes us all to remember who had that vision first.

While working for Xerox my first ten years out of college I got to meet a lot of interesting people, and see some absolutely *astonishing* technology. I learned the Xerox Star workstation when it first came out, and then trained people on its use. Our department had an Alto machine. I learned Smalltalk on that Alto. The wonder of Smalltalk wasn't in the language so much as in the interactive development environment that came with it, an environment that foreshadowed Visual Basic, Delphi, Lazarus, and scores of other multiwindow editors and GUI builders. When I sat down in front of our Alto, I smelled the future—and that future eventually came to pass.

Our Lady of the Analyst Guide

From Visual Developer Magazine #50, June/July/August 1998

Earlier this year IBM sent me a copy of their Via Voice Continuous Speech Recognition package. Hey, man this is what science fiction used to be about! And since I had all these old SF stories in a drawer that I had written (on an IBM typewriter) long before the era of personal computing, I figured it would be apropos to dictate my old SF onto disk using this amazing new IBM technology. Well…sometimes you don't have to be funny. Sometimes the material does it for you. Without further ado, here's first the start of what I wrote—and then what ViaVoice heard. Forgive me, IBM, but the future isn't *quite* here yet.

Our Lady of the Endless Sky

By Jeff Duntemann

Under a glassy dome made invisible by the lunar night, the Mother of God stretched out her hands to embrace the stone horizon. Beyond the tips of her marble fingers rock and steel lay ash-gray under a waxing Earth. Above her peaceful white brow the stars stood guard to all eternity in a sky so deep it had no bottom.

In front of the native granite pastels in the nearly finished church, Father Bensmiller knelt and prayed.

Let them see what I see now, Mother, and they would run to you.

A faint crunching vibration entered his knees from the dusty floor, newly enlaid with pastel blue tile. Bensmiller looked up. Bright light flashes off metal dazzled his eyes. The polished aluminum boom of a crane hove into view and wobbled slowly out of sight beyond the wall that supported the transparent dome. They were driving it to the construction site, where a third of the station personnel were planting new machines in the lunar soil.

Bensmiller went back to his personal miseries at the feet of the statue. Not an hour before, Monsignor Garif had spoken to him on the

S-band from Houston. As twice in the past, the news was of the rising number of American churches closing their doors permanently. Not due to a lack of funds; the Interfaith Council assured each pastor a living and attempted to keep the buildings standing. It seemed pointless, however to preach the Gospel to empty pews.

Our Lady of the analyst guide

By Jeff Dunn to the

Under glass the deal made invisible by the lunar night, the mother of God stretched out her arms to embrace the stone horizon. Beyond the tips of their marble fingers rot in steel late-grey under all waxing the earth above her peaceful White Ball the star stood guard to Wally turn a day in a sky so deep it had no bottom.

In front of the native granite pedestal in the nearly finished Church, Father Benz Miller's Delta and pray.

Let them see what I see now, mother, and they would run to you.

A faint printing vibration entered his knees from the dusty floor, New Orleans and laid with pastoral blue tile. Dennis Miller looked out. Bright light-flashes of metal dazzle his eyes. The polished aluminum pool more of a crane holes into view and while slowly out of sight to be on the wall that supported the trans. Dome. They were driving it to the construction site, where a third of the station personnel were planting new machines in the lunar soil.

Then as Miller went back to his personal misery is at the feet of the statute. Not an hour before mine senior dear reef had spoken to him on the S band from Houston as twice in the past the news was of the rising number of American churches closing their doors personally not due to lack of funds; the interface Council assured each pastoral living and attempted to keep the building standing. It seemed. Lists, however, to preach the gospel to empty perfumes.

If U Cn Rd Ths...

From *Visual Developer Magazine* #53, January/February 1999

People sometimes ask me why I don't have my PGP public key in my email signature. After all, I've written about encryption in these pages many times. Surely I don't send my email *naked* across a public network!

Well, um, yes I do. Truth is, I almost never put anything in email that I wouldn't mind seeing on the front page of the *New York Times*. I'm not any sort of activist, I'm uninterested in drugs, I love my wife, I don't frolic with interns, and I don't cheat on my taxes. Basically, I'm a boring guy.

But there's another reason: Data encryption isn't the same as privacy. It's a big step, fersure, but incomplete. (Hey, if I send a strongly encrypted message to a Caribbean bank, the government will have some cause to assume I'm up to something...) There are actually three issues in communications privacy: #1 Who I am. #2 Whom I'm speaking to. #3 What I'm saying. Encryption only covers #3, but in fact a great deal can be inferred from #1 and #2.

Anonymous remailers claim to cloak sender identity, but they're only as secure as the server hosting the remailer program. The Scientologists went all the way to the Finnish government to crack penet.fi, and it wasn't that hard to do. Security that depends on government self-restraint is no security at all, heh.

I haven't yet figured out how to cloak sender identity, but I have an interesting suggestion as to how we might prevent anyone from determining who we're sending messages to. It depends on encryption, but it also depends on bandwidth and processor power, so it's really a mechanism for 2010 and not 1999. Still, once the country is heavily laced with multigigabit fiber, this sort of thing will work.

Consider a Usenet newsgroup. Anybody can post to it, and anybody can read it. Posting anonymously to a news group is tough—anybody who really wants to trace a posting can trace it, even if you dummy out your name and Net address. However, tracking who reads a newsgroup, while theoretically possible, is orders of magnitude harder, because a newsgroup doesn't live in just one place. A newsgroup is replicated at any number of servers around the world, each of which is a separate point of access.

So...imagine with me a future scenario where people post strongly encrypted messages to a server using the recipient's public key. This server is something like a news server, in that all messages are accessible freely to all. But they're intended for robots, not humans. Robotic clients suck down all the messages and apply their owners' private keys to each message, discarding a message once it's obvious that it wasn't encrypted with their owners' public keys. Tens of thousands or even millions of messages might pass through such a robot's sight before it snags one addressed to the robot's owner.

This is why it's a technique for the future, when processor power and bandwidth are much more abundant than now. Keep in mind that typical text messages (like those executing funds transfers) are quite small and can be compressed almost out of sight. A 2010 decryption robot might be able to process a million messages or more per hour, and multigigabit fiber networks could transmit a million or more such messages per *second*. 12G hard drives are $250 today. What sort of storage capacity will servers have in 2010?

In short, you post encrypted messages in thousands of public places and then monitor those places for messages encrypted for you. If you can read it, it's yours. If some other guy can read it, it's his. The use of bandwidth and CPU time are not a waste, really, but the price of privacy. And it is privacy, finally. After all, the more people use the system, the less the government can infer from it. If use of the system becomes universal, I suspect that it would be uncrackable.

At that point, I'll start publishing my public key. In the meantime, I'll just be boring, OK?

For a long time I assumed this would never come to pass. And maybe it hasn't, even by now in the early days of the year 2021. But back circa 2005, I got on Usenet again, after not having had an account since my Internet provider stopped providing Usenet in the mid-1990s.

What I found after 10 years' absence was that Usenet was mostly empty except for binaries; i.e., files. Most of the cool and crazy conversations I used to have on groups like alt.life.afterlife were gone. In their place were endless numbers of files of all kinds: MP3s, images, and executables, most of which were malware. I was looking at it less and less, until I canceled my account as pointless circa 2017. However, not long

59

before I ditched the account, parties unknown were uploading binary files encoded as text. Lots of them. There was no clue what the files were about. I downloaded one and poked at it, but I'm not a codebreaker and had better things to do. I have two theories about those files:

1. They were encrypted backups, stored on Usenet as though Usenet were another species of cloud—and with binary retention of three years or more, it is. Or

2. They were encoded messages of the sort I proposed in this idea piece.

I still don't know, and it doesn't particularly bother me, since the possibility remains: Some bright person could implement such a system. We've got plenty of cycles and bandwidth—more, in fact, than such a system would require. So maybe it *does* exist, and just keeps a low profile. Given what strong encryption is often used for, it had *better*.

Dear Sir: Your Lumber Is Ready

From *Visual Developer Magazine #57*, September/October 1999

In 1980 Sweden's Department of Forestry wrote a letter to the Swedish Navy, informing the Navy that its lumber was ready. One hundred fifty years earlier, in 1829, Swedish planners predicted that a shortage of oak timber suitable for building warships would arise by the year 1990. So they instructed the foresters to be proactive, and the foresters planted a new oak forest on a government-owned island, knowing that it takes ship-quality timber at least 150 years to mature. In their view, I'm sure, they felt that they had barely forestalled certain disaster.

Stewart Brand tells this wonderful story in his recent book, *The Clock of the Long Now*. The book is a plea for a return to the long view in human society, like the one the Swedes had in 1829. As intriguing as the book is, Brand tiptoes around the question of what exactly constitutes a long view, and while he credits the Swedes for long-term thinking, he's at a loss to suggest what they should have done. Nobody can predict technology—nor anything else, for that matter—one hundred fifty years in the future. So how do you foster a long view of *anything?* What the Swedes probably should have done is funded a think tank on the future of warfare. On the other hand, the oak trees were a good hedge, and I guess you can always make bookcases out of them.

To take the long view is to think and hedge, perhaps—remembering which is which. Optimism is always called for, though it should stop short of mania. It should be obvious that long-view endeavors must produce more than they consume. (This is no longer obvious to many in our society, as I learned in reading Michael Wolff's acerbic *Burn Rate*.) But for my money, what marks long-view thinking is one commandment more than any other: *Keep your options open.* Dead ends aren't just limitations in space. They mark an end point in time, and the premature end of progress.

Long-view thinking, in technology especially, is mostly the avoidance of limits. Let me offer three points to ponder in the pursuit of the long view:

1. Algorithms often outlive their implementations. Don't let your implementations infect your algorithms and thereby limit them. (This is what Don Knuth was getting at when he said, "Premature optimization is the root of all evil.") Y2K is the idiot's example here, but subtler examples are lodged everywhere in our code like buckshot in a burglar's behind. You'd think we'd have learned better by now…

2. *Openness is less limiting than secrecy*, which, taken to an extreme, is *forgetting*—the destruction of information and hence of options. An open source product has many more options than a proprietary one, and will live longer.

3. *Develop a sense of wonder, and incorporate it into your planning*. Doers think. Leaders think big. Earth-shakers think *wild*. History tells me that over the long haul, wild trumps big every time. Put your longest view into your *ideas*—somebody will eventually be able to come up with an implementation.

Finally, implicit in the very idea of a long view is that there is such a thing as the common good, which both outlasts and is ultimately more important than any individual, group, or nation. Adopting rather than subverting existing standards is a long-view tactic, as is the support of open markets. (The freest economy in the world—ours—is also the strongest. This is no coincidence.)

As for the Swedes, well, their lumber is ready. The dumb response would be to harvest the trees and make bookshelves out of them at 3% retail margin. The smart response would be to keep the forest. The brilliant response would be to build yachts, sell them, and then use the proceeds to plant ten times more trees, and leave the question of how best to use all that good oak to those who will harvest it in the year 2150.

Back when it was published, a friend objected to this piece on what seemed at first to be solid ground: Taking a truly long view requires predicting the future. My reaction is…so? The infamous memory ceilings we ran into on a regular basis during the first decades of computing could have been prevented by some very solid predictions of the future: that semiconductor memory would grow cheaper and cheaper as years passed. Anybody who knew anything about semiconductor fabrication could have made this prediction in 1980 with absolute confidence that it would be true. Granted, that's not been true for 150 years yet. And in fairness to our computing forbears, memory was once *extremely* expensive, which in part excuses the Y2K issue. Having taken a long view, they reasonably supposed that software written in 1975 would no longer be used in 2000, so they could cut memory corners in 1975 without consequences. *Surprise!* So a cost-benefit compromise shocked everybody by living what, in computer terms, is forever.

There's a flipside to the long-view problem: Taking a long view…and then not capitalizing on it. The kings in that castle were the suits running my former employer Xerox, who threw money at PARC and commanded them to invent the future of computing. PARC did it. They were right. (I used an Alto in 1983. Windowed environments? Mice? OOP? Yup.) And then the suits threw it all away and tried to sell Z80 CP/M boxes two years after the IBM PC took over.

Keep your options open. That's the best long view of all.

Jeff's Singularity Rant

From Visual Developer Magazine #57, September/October 1999

I have a middle-aged writer friend (who lives somewhere in the midwest) with a problem. For the past four years, something has been swiping his pain pills, coins, paper money, and even whole bottles of wine. The "disappeared" items come back at erratic intervals, often materializing out of thin air in front of the poor guy's face. On unusually active evenings, pennies will appear out of nowhere and fly across the room, and packages of Tootsie Rolls will explode, scattering little brown cylinders all over the kitchen. He's been meticulously documenting these mysterious goings-on in a private journal, trying to make some kind of sense of it all. So far he's failed.

He's tried to interest scientists in the phenomenon, to no avail, because nothing ever happens when anybody else is around. And, of course, he's taken a beating from "skeptics" who try to convince him that such things are impossible and he's just imagining them. He's pretty much given up trying to convince the world that something funny really is going on, and I'm convinced that he remains sane mainly because his friends (including me) keep telling him that he is.

I mention Tom here because I am profoundly irritated at the Extropians and their ilk, who are lead-pipe certain that all the advancement-of-technology curves are heading for their exponential knees, and that sometime on or about 2012 (which is when the Mayan Calendar ends—could it be a coincidence?) computer power, knowledge, bandwidth, and consumer demand will simultaneously become infinite and we will all vanish into some ill-defined teleological Black Hole.

This is the Doctrine of the Singularity, and I (who have been lumped in with the Extropians in some quarters because of my profound optimism—which, I hasten to add, stops short of resembling a cocaine frenzy) just gotta smile. The truth is, the Singularity exists by virtue of extrapolating advances in engineering (and only certainly carefully selected branches of engineering, too) past the point where the curves themselves become absurd. Engineering always has limits, and we are currently chasing the limits of semiconductor technology. Once FET gate insulator thickness gets down to five or six atoms, we'll be there—and so will the rest of the world, Singularity notwithstanding.

If you're looking for potential singularities—which when the cocaine-frenzy mumbo-jumbo is subtracted, cook down to world-wrenching changes in the human condition—you need to look not at engineering, but at basic science. And when I look at basic science, I fail to see many world-shaking discoveries in the last fifty years. The wild ride we're on now is the tail end of our mastery of solid-state electronics, which began in 1947 with the invention of the transistor. Computing as an idea actually predates the transistor, and after 1947—well, to me it's been small stuff indeed. To get on the next engineering roller coaster, we need some genuine sense-of-wonder stuff out of the scientific community, and I have yet to see anything that beats the humble—and world-wrenching—transistor. Even quantum logic is just a special, clever kind of transistor. Engineering. Not science.

You can laugh at the notion of telekinesis and teleportation—what my friend Tom is basically dealing with—until somebody learns to control it. Then, *duck*. I am ashamed to report that many scientists claim that ball lightning doesn't exist, because it can't be explained or recreated on demand. And yet ball lightning came down her farmhouse chimney and nearly killed my mother in 1936, which leaves me with a *highly* personal stake in the question. On the surface of it, ball lightning seems to draw upon some unknown source of energy—and that's gotta be at least as useful as talking Furby dolls, right?

The Extropians doth protest too much, and a singularity may simply be the blank wall that rears up when progress stalls out. We may face such a blank wall in 2012 or so, unless we can figure out how to aim the searching eye of science in new directions. Telekinesis/teleportation and infinite free energy could break that wall. Oh, yeah. Telekinesis and infinite free energy are impossible. Sorry. I keep forgetting. So...how 'bout that new Nintendo?

I brought Tom's poltergeist into this essay because the weirdness he's experienced could be evidence of new physics that we haven't yet touched—because we're so sure that they're impossible. As for Tom, I've lost touch with him. I should ping him. As some readers of this piece have accused, it's possible that he's making it all up. But...why would he? So that 99% of the world would call him crazy? Wouldn't be my first choice either. He's writing in his journal and keeping quiet. That suggests he's experiencing what he claims. Poltergeists aren't rare, as Colin Wilson has documented down the years. I'm keeping an open mind.

Hail the Millennium!

From *Visual Developer Magazine* #58, November/December 1999

I was six in 1958, and with Explorer I in orbit on the heels of Sputnik, the world was drunk on the future. "Closer Than We Think" ran in the Sunday funnies, and I devoured every library book I could find with pictures of spaceships, satellites, and monstrous machines that ground across the landscape, leaving gleaming miles of interstate highway in their wake. The future was a *place*, and I wanted to live there. Why? Because it was Better.

Perhaps the oldest myth to animate the human spirit is that tomorrow will be better. It's what pulled us out of our caves and then our huts and then our grimy, overcrowded cities. For ten thousand years we've relentlessly invented and revised and scolded and reformed, and inch by inch the human condition has improved. Against this progress has stood the same sad refrain sung by the same damned cynics, complaining that everything is a gradual descent into perdition from some ill-defined Golden Age.

Don't buy it. Back in that Golden Age, starvation was the norm, and we practiced human sacrifice and systematic murder and genocide on a scale that makes us all here today look like Mother Teresa. Ignorance, paranoia, and hatred were a sort of world religion. And it's not just the Dark Ages of which I speak. The best I can say for the Fifties is that we were a little more polite to one another—but beneath that brittle shell it was a time of seething hatreds and horrible injustice.

I say all of this because I read history, something almost nobody else seems to do anymore. I also try to keep our expectations in perspective. My solidly middle class great-grandfather thought he was doing well to have running water. Today, even the poorest of American homes has a color TV and a telephone. There remains plenty of poverty on Earth, but it's a far gentler poverty than it was even a hundred years ago, and more people live more comfortably than ever before.

So are we there? Is it really Better today? I think so, not only for the obvious improvement in physical living conditions brought about by technology, but also because *we know what the real problems are*. It's begun to dawn on ordinary people that what makes us miserable are spiritual hungers and lacks, not physical ones. This is a tremendous victory, and an awesome and underappreciated insight: *The forces that impede our happiness are no longer imposed from without, but come from within.* Technology has laid the physical world at our feet. We are freer, wealthier, and better educated than at any time in history. If happiness eludes us now, it's nobody's fault but our own.

On December 31, 1999, I will declare that I have finally made it to the future. That night I'll invite my friends over to barbecue, and Carol and I will waltz in the dirt driveway between the cactus and laugh and play corny songs and drink pina coladas and slap a lot of backs. We'll make a bonfire and burn all of our Ed Yourdon books, and perhaps stand for a moment and remember those who believed in the future but didn't live long enough to see it: My father, Uncle Louie, and all the unnamed people who like them helped nerdy young boys believe in themselves and the future too, and in doing so helped make it all happen.

For it really is better now, as we would realize if we would read history more and stop the continual upward revision of our definition of misery. Few of us have all we want. But most of us have all we need. We have achieved the future. We're pointed in the right direction. The rest is—dare I say it?—spiritual engineering.

So bring it on, all of it! Hail the Millennium! The next thousand years are gonna *rock!*

We had that party on December 31, 1999. It was wonderful. We had authors and editors and sales reps and artists and framers and engineers and programmers and a lot of kids. We didn't burn any books; Ed Yourdon's tomes had gone in the trash some weeks before, and he's off my radar now forever. We ate, we drank, we talked with our friends, we played with the dogs, we remembered our dead, and set our sights higher than the latest political shitshow. Courage and a determination not to harm others will see us through into the future. It's worked before. It will work again.

Three Millennial Challenges

From *Visual Developer Magazine* #59, January/February 2000

Well, we've put Y2K behind us, and it was an excellent excuse to dump thirty years' worth of crappy software and obsolete iron. (It was also a great excuse to buy a new gas grill and a solar panel, which I've always wanted.) What's next? Are we in the clear now?

Not quite. Computing has solved most of its problems, and the solutions to others (like fat pipes) seem distant only because we're impatient. I may even get DSL here next year, sheesh. Cycles, RAM, and disk are abundant and cheap. Win2K and Linux are terrific platforms, and who knows? The Mac may survive its perpetual dance on the edge of the abyss once Apple's management grows up and stops being jerks. There remain three pretty dire issues before us, however, to which solutions are desperately needed but are nowhere in sight.

1. *Malicious execution.* Some aspects of this, like Word macro and JavaScript viruses, are idiots' work, and only await somebody with enough guts to create Web browsers, email clients, and word processors that draw the line and say *content will not execute.* Preventing more traditional .EXE file viruses will require that operating systems forbid executable files to be modified once installed—even by themselves. Technically Microsoft could solve this problem, but they dare not, because to solve it they must first admit that it's *their* problem, which would bring on the lawyers like piranhas. The more losses people suffer from viruses, the greater Microsoft's reluctance will become.

2. *Language comprehension.* Speech recognition, while gnarly, is a solvable problem. It's language that's the kicker. Yet without good language comprehension, we've pretty much gone as far as we can in software usability. The mouse wars are over, and we're down to fiddling with screen widgets now. Further breakthroughs will require software that understands context and culture. Researchers are beginning to admit that the problem of language is the problem of how the human mind works, something that many, myself included, consider unknowable. To make any progress at all we will have to take a coldly objective look at what human culture is and why people do what they do, which is something no one seems brave enough to attempt these days.

3. *Privacy.* The worst problems we face aren't technical at all—and that's why they're so scary. Privacy is a force that pulls in two directions at once: your privacy, after all, is my anonymous harassment. To eliminate spam we have to eliminate anonymous email, and the technology community is split down the middle on that one. The government fears encryption not because it fears crime or terrorism, but because it knows that electronic anonymity will eventually make income taxes—and thus government as we know it—impossible. That's why no one will bother with e-cash unless it's two-way anonymous and provably unbreakable, because the sole purpose of any kind of cash these days is to evade income taxes. We already have *traceable* e-cash—it's called your Visa card.

Looming over all of these is a shadowy fourth problem: Government. When computing was locked up in a glass-walled room, nobody but paranoids considered it a threat. PCs and networking put computing in everyone's face, and now everybody—feminists, fundamentalists, record labels, software companies, movie studios, whoever—wants government to address their private issues to *their* benefit, and to hell with everyone else and the common good. This comes at a time when big money and an electorate divided almost surgically in half have paralyzed two-party democracy, giving us *de facto* rule by lawyers and judges.

Eek! Makes Y2K look almost cuddly, doesn't it? And although there are times when I sympathize with Dick the Butcher, it's dangerous to stand outside the issues and point fingers. We need to keep working on solutions, certain that solutions will eventually turn up, even if they're not the solutions we envisioned. If post-WWII history has taught me anything, it's that things do work out once people assume that progress can be made. We don't always get what we want, but we generally get what we need. Trust the emergent future—keep your hand in—and always be ready to be surprised.

I think I nailed this one to the wall. The three problems I cite here are still with us—and more urgent than ever. We've made progress on all of these problems except the last one. Government is gnarlier than ever, especially with Congress, the Senate, the Presidency, the Media, and education now under the almost complete control of one political party. Maybe that isn't a problem. We won't know for a couple of years. But one reason I haven't bought cryptocurrency is that I think we're looking at the eventual outlawing of anonymous digital cash. People are buying things with crypto, and have been for a few years now. It's when ordinary people begin to get *paid* in crypto that the Feds will bring the hammer down. That's not the kind of attention I want to attract.

As for the first three, we have a handle on malicious execution, at least from a technology standpoint. What problems remain are mostly human ones. Social engineering is a thing, and an awful lot of people just can't bring themselves to do some research before clicking on a link or (especially) installing software from an untrusted source. Language comprehension will come in the reasonably near future, depending on how we define "comprehension." Computers already understand context sufficiently well to be useful under (simple) voice command. I ask my phone questions all the time, and the answers I get continue to surprise me with their pertinence to what I asked. But that isn't "thinking."

There's some pushback against privacy violation in social media, which didn't exist as a force when I wrote the original essay in 2000. Our privacy problems have a lot to do with "free" as a business model. If you don't pay for the product, you *are* the product. How social media pays its bills without selling your life details will be the big story of the new Roaring Twenties. What I would like to see is some sort of law forbidding user tracking altogether. Not ads, of course. Just tracking. Could social networks sell enough ads without tracking demographics to continue operating? Nobody knows, and I doubt the experiment will ever be done. I do think that the biggest social networks have now pissed off enough powerful people on both side of the political aisle that there will be legislation to limit their power.

Then there's the insane way that the biggest social networks have decided that certain voices (nearly all of them conservative) *will simply not be heard*—without any explanation as to why. I don't think any of us expected back in 2000 that something like a fancy computer bulletin board system would have anything like the power that they now have. But that may be the biggest single problem in the computing universe now, in 2021. I have some ideas on how that might be fixed. I just don't have a magazine to publish those ideas in anymore.

--73-- --JD--

From *Visual Developer Magazine* #60, March/April 2000

The Final Issue

Those of you who know me well (and it's amazing how well some of you seem to know me!) are aware that I am an unshakable optimist, that I love technology, and believe in the future with a sort of crackpot manic intensity usually found only in religious fundamentalists. So it's a little odd to ponder that the same technology I've been cheering in these pages has gradually eroded the foundations of this magazine, and the day I've been dreading for over a year has finally arrived: With this issue we are laying *Visual Developer Magazine* to rest.

The problem we've had in keeping the magazine viable is an interesting one. As the years have passed, developer technology has grown frighteningly complex and interconnected. Ten years ago it was DOS and conventional programming languages. How much space does it take to explain procedural parameters in Pascal? You can lay the whole thing out in five pages and assume the reader won't go home hungry. Now, fast forward to 2000 and in those same five pages teach me all about COM objects and Delphi. When you stop laughing you'll see that that's not a magazine article. That's a whole book. (VDM regular Eric Harmon has just written the book: *Delphi COM Programming*, from Macmillan. Fine stuff; snag it!) Basically, to cover modern developer technology usefully takes more room than a printed magazine can afford. The smaller scraps that remain are so audience-specific that they can't float an advertiser base, and these days can be had for free on the Web.

I know it's not just us, because other advanced-topics computer magazines arc falling right and left. This is our sixtieth issue, representing a ten-year run. Not bad at all. So rather than become a gossip mag or publish nothing but product reviews, we decided to gracefully accept the dictates of evolution and admit that our mission can no longer be achieved in magazine format. We have struck a deal with online publisher EarthWeb to acquire our circulation liability, and those of you who have paid balances will shortly receive a letter from Earth-Web, offering you either a pro-rata cash refund for unserved issues or a one-year subscription to their ITKnowledge online reference site, which ordinarily costs $195. That sounds like a pretty good deal to me.

Many of the people you've come to know in VDM publish elsewhere, online and off: Al Williams is an editor and columnist for *Web Techniques* (www.wcb-techniques.com) and publishes in all sorts of other places too. Eric Harmon is entering a new career writing Delphi books. (Hurray!) Matt Hart has a superb site on advanced VB topics: http://blackbeltvb.com. David Gerrold has a column in the new Galaxy Online SF Web site (www.galaxyonline.com) and shortly I will too. Ray Konopka tells me he will be updating his Delphi components book for Kylix (Delphi for Linux) and is looking for a new home for his column. All who worked on the magazine here at Coriolis have new jobs.

What can I say in closing? I have this intuition that the work we're doing in computing today, no matter how mundane or pointless it may seem at times, is the primordial foundation of something wonderful, something that in the decades and centuries to come will be seen as the very heart of technology and indeed all human endeavor. In a manner not unlike Tolkien's First Age of Middle Earth, we will be seen in the year 3000 as creatures almost mythic, walking around in a landscape illuminated by the strange glow that comes of seeing things happen for the very first time ever. *These are the people who invented computing,* they will say, and they will imagine our lives with awe.

I take some comfort in that when my eyes get tired and my back hurts. That, and the signature statement of my patron saint, Lady Julian of Norwich: *All will be well. And all will be well. All manner of thing will be well.*

It has been an honor to serve you. If it means anything at all, I believe in you. Now let's get on with building the future of humanity, one line of code at a time.

This was one of the hardest essays I've ever had to write, for obvious reasons. *PC Techniques/Visual Developer* was the finest thing I have ever created. Those were good times. I learned a lot, I met a lot of brilliant people, and helped create technical content that folks still talk about. I wonder sometimes what I'll be remembered for after I hang up my meat-suit. The Coriolis Group, our books, and our magazines would definitely be on the top of a pretty short list.

RAD Mars

From *Visual Developer Magazine* #60, March/April 2000

The Final Issue

Whew! We ducked the imaginary Y2K bug. (*All we have to fear is fear itself*—are you sure that guy didn't run Linux?) So what's next? I know! Let's model a planet—in 1:1 scale!

We have meter-resolution elevation data for virtually the entire surface of Mars. Let's start with that, and design a rendering engine that either interpolates and textures the meter-scale points realistically, or uses fictional points created by designers (us) that render down to the centimeter level. Pass the RADMars server a set of Mars global coordinates, and your Web browser looks out on… Mars.

Physics works like it does here, adjusted for Mars conditions: Throw a rock, and it goes…far. Edge off a cliff, and you fall—and when you hit the bottom, you "die" and wake up back at Mars Port Alpha. The Sun rises and sets. The wind is thin, but it registers on your "suit" instruments in your browser status bar.

It's the ultimate collaborative Internet portal. For we won't stop at simply gaping at Marscapes. We get it working, and then we build. Part of the RAD Mars design is an ErectorSet API. There's a catalog of parts—girders, dome segments, panels, windows, maglev track sections, even architectural fittings down to drawer pulls and sconce lamps—and with those we begin to create houses, towns, community spaces, and entertainment facilities. (Your construction equipment is a browser plug-in.) If you've got a rock band, design your avatars and then book a time slot to come down and jam. Sell your music CDs to all the other avatars who show up to listen. Want your avatar to breathe fire? There's a function call for that in the avatar API.

And who's "we"? Why not *everybody*? There's a *lot* of Mars. 56 million square miles of it—more than enough to give an acre to every human being on Earth. (There are 640 acres in every square mile. Do the math.) We can set up a homesteading program. Register to play, and you get your acre. People and companies that want more can buy acreage from the portal to defray portal expenses.

Whatever's inside your acre is your business. (Sound from your rock band conveniently stops at a "force field" on your property lines—which also keeps out all visitors but those you explicitly invite.) Public space is governed by committee. Artists, architects, and engineers could make money by designing new parts—even entire structures—and selling them. Being virtual makes certain things easier—like moving your Frank Lloyd Wright home to a different plot if somebody else wants to swap. And we could stretch physics to allow Niven-esque "stepping disk" teleporters to get from one point on Mars to another instantaneously.

The portal would offer varying levels of resolution to Marsholders depending on their connection bandwidth. At 33Kbps, well, Mars looks a lot like a Doom level. But at cable modem resolution, things start looking like *Myst* rendered in real time. And at the T3 or better bandwidth that optical connections will eventually give us, you're talking photorealistic animation. (Am I being too optimistic? Who knows? Look where we were ten years ago. Now imagine where we'll be ten years from now!)

With the rules and operational mechanisms set up correctly, RAD Mars could become a seething cauldron of new ideas in art, music, collaboration, and governance. If it became popular, it would drive development in fast graphics and especially fast Internet connections. It would certainly provide a mass market for exotic VR helmets and tactile feedback gadgetry, which has always been a solution in search of a problem.

Why Mars? Mars is a *world*, and more than that, a myth with a powerful hold on the human imagination. Mars invites that imagination to reach above the mundane limitations of meatspace life. The acre is free. The parts (the standard ones, at least) are free. The avatar is free. Everything, furthermore, is as malleable as your dreams.

Mars exists to draw forth our dreams. Let's go there before we forget how to dream entirely.

Perhaps appropriately, these words were the last words on the last page of the last issue of *Visual Developer* ever published. It was great fun, those ten years. Thanks to all of you who were there while it was happening, whatever role you played. You'll never know how much I appreciate it.

PART 2. From Contrapositive Diary

I invented blogging. Did you know that? Well, sure, a lot of other people invented it too. I mean, how obvious could it be: Write something every day (or every so often) with a date and a title. It had been done in 1994, if the accounts I've read are accurate. I didn't see those early blogs. I thought it would be an entirely new thing.

Ok, I didn't invent blogging. Or let's say the original idea came from somebody else: My ad sales rep for *PC Techniques/Visual Developer*, Lisa Marie Hafeli. Lisa thought it would be a cool idea for me to write something every day on the Coriolis Group website, as kind of an online daily diary. My advertisers liked what I wrote, and thought that my writing was a big draw in the magazine. So, could I put something up on the Web every day, and maybe work in a few product mentions, hint hint?

I sat down with Dave, the Coriolis webmaster, and he thought it was a great idea. So, starting on June 5, 1998, (almost) every day I emailed Dave a short block of text, which he posted on our site the next morning. It's a little unclear how many more ads we sold because of VDM Diary, but I found myself enjoying it so much that after we laid the magazine to rest in the spring of 2000, I bought a Web hosting account and kept going, under the name Jeff Duntemann's Contrapositive Diary. I broadened the scope of the diary to…everything. Obviously, computing was a big part of it, but I wrote here and there on almost everything that interests me.

Contra is hardly a daily thing, but since 1998 I've posted over four thousand entries. In this section, I'll be publishing some entries that have withstood the test of time. Or maybe the ones I just liked the most. All entries are online, going all the way back to the original one on 6/5/1998. It's been fun, though a lot of work, and in these turbulent times I'm trying to avoid provocative topics. But it's as close as I've been able to come to the sort of fun I had editing *PC Techniques/Visual Developer*. Take a look. If you like what you see, well, there's a dumptruck more online.

50 Days' Meditation on Writing

Originally posted on Facebook in 2014

1. It all starts here: *Read*. Read voraciously. Read like your head is starving, because it is. The machinery inside you that creates feeds on words. Unless you stoke that furnace, what will come out (if anything) will be stale and cold.

2. Write. Write every chance you get, whether you feel like writing or not. It doesn't matter as much what you write as that you're putting the engine in gear. If nothing's happening on a formal project, well, just joyride at your keyboard. Think of it as inviting your subconscious mind up for a beer. You may be surprised at what comes out.

3. Read outside your primary interests. I shorted psychology, history, and theology as a young man because I found science and technology intoxicating and had a natural talent for them. Creating characters was thus secondary to creating universes, a problem I didn't even know I had until I was almost middle-aged. I knew I needed characters, but I created them by imitation because I had no idea how to build them intelligently. That worked in the pulp era. It doesn't work anymore.

4. Read the sorts of fiction you don't like, especially genres that are doing well. Read them with an eye toward figuring out how they work. What do those guys on the other side of the Mysteries hill know that you don't? This doesn't mean you have to start writing mysteries. What it means is that you have to be willing to eat a little kale to know what kale is. And who knows? Maybe you'll start liking kale.

5. Read weird stuff, especially if you want to write fantastic fiction. You don't have to actually *believe* in bigfoot or UFOs or ghosts or dowsing or banshees to get a sense for them. Plus, the people who do believe in them and write about them are often wonderful archetypes for eccentric characters.

6. Live richly and consciously. Fiction that doesn't grow out of experience is just imitation. Travel, snorkel, ski, climb rocks, cook, dance, touch dinosaur bones and living stalks of wheat. Go to zoos, look at the stars, smell not only the cinnamon bread but also decay, disease, smoke, and grease. Buy some tools

and build something. Find a worthless artifact and pull it apart, because there's something inside that may be worth seeing, something that may prove invaluable later on. Whatever you end up doing, pay attention to what you do and how you do it. Living on automatic gains you *nothing*.

7. Look for the causes and effects that drive life forward. A lot of the art of fiction lies in showing people how one thing follows from another, and how nothing just sits on a rack above the fireplace forever.

8. Look for the connections between things. This was one of my "super powers" and may have been an inborn talent, but it's a skill that most of us have and may be honed with practice. Go watch some James Burke—scratch that, watch the whole series. Every so often a connection you discover will be startling. Write it down.

9. Pay attention to the minor senses. The eponymous heroine of *Amélie* (2001) enjoyed thrusting her hand into buckets of grain. This puzzled me until I ran across a bucket of grain. Then I understood.

10. Dare I say it?—go to church. Whether God exists or not, there is something in the depths of the human spirit that you will never taste unless you allow yourself to be open to the transcendent.

11. This trick served me well as a teen: Type a passage from someone else's story that you find compelling into your keyboard, and stare at it for awhile. Don't lift it from an ebook or online discussion. *Type it*. The act of typing it allows you to perceive and appreciate its makeup *as a writer*. You'll spot things like clever dialog and paragraph structure that you would not have noticed while simply reading it.

12. Critique from readers may in fact be more valuable than critique from other writers. Readers know what they want, what they enjoy, what moves and astonishes them. Writer friends have a tendency to focus on the how. Readers focus on the what. You need to pay attention to both.

13. Be aware of your habits, and deliberately subvert them. Much of this is a willingness to operate outside your comfort zone. Ruts are just horizons pulled in too close, after all. Push 'em back. If need be, find broader horizons.

14. Sing. You don't have to join a choir. I have malformed sinuses and my voice isn't terrific, but I sing in the car while driving alone. Why? There's something a little bit mysterious about sung words. Different paths within the brain are en-

gaged by song. People suffering from certain forms of aphasia can't speak…but can sing. I've gotten some startling insights and scenes while belting out banal pop songs from the 60s at the top of my lungs. Try it.

15. When you get stuck, don't push. Let go. Writing by peeing waterfalls hurts less than writing by shitting bricks.

16. This is an ancient proofreader's trick: You can often spot typos in your own work more easily by starting at the last word on a page and reading the page backwards.

17. Vocabulary matters. Every time you look up a word in the dictionary, read the whole page.

18. Another trick I use with a massive 3-volume dictionary from the 1930s is this: Flip through the dictionary's pages, glancing at the little drawings. (Older dictionaries have lots of them.) If you see one that's unfamiliar, read its entry.

19. Special-purpose dictionaries make good bathroom reading. *A Concise Catholic Dicionary* is a good example, especially if you can't tell a dalmatic from a tabula. If you know what something looks like but not what it's called, you can't write about it. DK's *Eyewitness Visual Dictionaries* series is wonderful for this problem, as is *The Macmillan Visual Dictionary*. Others exist, on a broad range of topics. Find some on the used market, and let them become the power behind the throne.

20. Writing is *not* a zero-sum game, and neither is publishing. (Well, ok, publishing is to some extent, at least for the time being.) Encourage your fellow writers. Don't fear them nor be envious of them. No one but them can write what they're writing. No one but you can write what *you're* writing. Publishing is increasingly about becoming well-known *outside* of traditional publishing. Work on your brand. Help your friends work on theirs. Cooperation helps keep competition from becoming predation.

21. Don't fall for stupid fads. Eliminating all adverbs that end in "ly" from your work is idiotic, and doubly so if you don't know what adverbs are, nor how they should best be used. People will laugh at you if you start railing against horrible words like "only" and "early." (I'll be among them.)

22. Theodore Sturgeon himself told us at Clarion: "If you're going to be an SF writer, don't be a writer in your day job." Yup. When that well goes dry, it's dry. I got stung by this bigtime when I left technology in 1985 and went into techni-

cal publishing. My fiction output fell to zero and stayed there for ten or twelve years. It was only when I got kicked upstairs into management circa 1997 that the stories began to emerge again.

23. One reason I blog less than I did five or six years ago is that I'm trying to get serious about fiction again, after a long dry spell. So where did all that dry come from? I have suspicions, heh. Practice is good, but ideally it should be practice at what you're trying to produce. Blogging and social networks like Facebook don't help much here because they're just too easy. Write uphill for practice and production. Write downhill for release.

24. I write my first drafts with the word processor page configured to look something like a printed book page and not a manuscript page. This allows me to develop a visual "feel" for things like paragraph size and sentence length that don't come through as clearly in a page set up in standard double-spaced manuscript format. When the piece is done and ready to submit, I make a copy and reformat it in the expected manuscript style.

25. Make peace with guns. This not only includes reading about them, but extends to going to a range and firing them. Don't let political ideology throttle your curiosity. You don't have to become a gun hobbyist to experience them. Guns are present in all types of literature, and if you portray them ignorantly, people will notice and dial down your credibility.

26. Don't idealize the past. The past was, for the most part, *awful*. This is one of the most important reasons to read history. Putting the past in perspective is useful not only in writing historical fiction but also in any fiction that portrays the human condition in detail…which would be most of it

27. .Listen to everyday conversation and write your dialog accordingly. People don't speak in meticulously crafted, 53-word-long sentences. People rarely address one another by name unless they're trying to get the other person's attention, or warning of trouble. People almost never say things like, "Boy, Teddy, that corned beef hash was particularly good once you got past the burned parts at the edges but before you hit the puddle of grease beneath the center."

28. A very good way to get a lay grounding in science is to chase down Isaac Asimov's old essay collections, like *A View from a Height*. Some science has changed since he wrote them, but most of his topics are of things fundamental enough to be timeless, and many are about the historical evolution of science, which you just don't read much about anymore.

29. Several writers who don't know one another have told me privately: *Political fiction doesn't sell.* They're right. Politics is filth, and it defiles everything it touches. Don't let it touch your writing.

30. Don't try to write in twenty-minute snatches. It doesn't work. You need an evening (at very least) to get some momentum. If writing is important to you, set aside the time. If you can't set aside the time, you're probably not going to be a writer.

31. Keep everything you've ever attempted, even your failures, no matter how embarrassing they may seem. You don't have to show that material to other people. Failures and fragments can be warnings of what not to try, or mines for characters, ideas, settings, and plots. Review your trunk regularly to make sure you remember what's in it. Even if nothing useful pops up, it will invariably remind you of how far you've come.

32. Certain concepts take over a story, most notably time travel and explicit sex. Any story that includes time travel will be *about* time travel, because the concept changes absolutely everything about reality as we understand it. Any story that describes sex in any detail will be *about* sex, because that's all most readers will remember a month later.

33. The most effective antidote to writer's block that I've ever discovered is to work on your memoirs. The topic is always fun, and you already know the material. Talk about yourself for half an hour, then put it away and go back to your current project. *Boom!*

34. Foster a cinematic imagination. Don't imagine a scene as text. Imagine it as a movie, watch it a few times in your head, and then describe it. Once you've written the scene, read it back to yourself while imagining the scene, to make sure that the two are still in sync.

35. When one of your characters suddenly acts out of character, it's a sign that something's wrong. Either the character is growing or his/her role is changing. Or both. Your subconscious (probably) knows what it's doing. If it's a good character, let him/her grow. You can bend the plot to fit.

36. Every so often, a character grows so much that the character completely outgrows the role you've provided. Minor characters—even lowly spear carriers—sometimes "get legs." (My character Jamie Eigen from *The Cunning Blood* was originally a sounding board for Peter Novilio to talk to, and then became one of

the pivotal characters in the novel.) You may need to decide if such a character would be better off as a major character in an entirely different story.

37. Polemic is hard. And dull. And mostly unmarketable. Don't bother. The best way to convert is not to persuade but to subvert.

38. In talking to writers and seeing what's published, I've sensed tremendous tension between what writers want to write and what their readers want to read. I'm improving my writing skills in part by trying different approaches in telling different stories. If you didn't know, would you guess that the same guy who wrote *The Cunning Blood* also wrote *Ten Gentle Opportunities?* Or "Whale Meat?" Publishers want to Sell More Of What Sells (SMOWS) and they're one reason why there are so many volumes in the Interminable Sameness Saga. Resist. Autotypecasting is a terrible thing.

39. If you can write good fanfic, why bother writing fanfic? You're wasting your talents. Let your imagination loose in search of your own universes. You might be surprised at what turns up.

40. Most serious writer's block is really depression, and depression is a serious business. (Alas, writers tend to be depressives.) If you've always been an avid writer and suddenly you can't force yourself to write at all for a month or so, worse things may be wrong than writer's block. Get medical advice.

41. A writer should know a little about a lot, a lot about a little, and something about almost everything.

42. A corollary to yesterday's entry, and truer of writers than most folk: The moment you stop learning is the moment you start dying.

43. As you design your villains, keep something in mind: Evil people, cabals, or organizations don't think they're evil. To the contrary, they think they're doing good in service to the world, and are enraged when the world doesn't immediately see things their way and surrender. Mustache-twiddling bwa-ha-ha evil overlords are comic-book stuff. To conjure a *real* evil overlord, you need frustrated idealism and a rejected savior.

44. Random, wand-waving, any-damned-rabbit-out-of-any-damned-hat magic like you see in Harry Potter strikes me as dull, and the very epitome of *deus ex machina*. To be compelling, magic must be a *system*, with known or revealable limitations and internally consistent rules. Make it too easy, and you will drain all the tension from your story, most likely turning it to soap opera in the pro-

cess. See Larry Niven's Warlock stories for a good example of magic done well. My novel *Ten Gentle Opportunities* is my own attempt at systematic magic, and people I trust tell me it works.

45. While it's true that the best way to give a character some depth is to work out his/her backstory in detail, many writers make the mistake of trying to insert too much of that literal, textual backstory into the tale at hand. *You* need to know the backstory explicitly. Your readers should be able to discern that backstory from how the character thinks, feels, and acts. Anything else may be experienced as expository lumps, or (worse) padding.

46. Collaboration, when the process is working well, can be a pretty intimate business. It's not writing together so much as writing together *naked*. Be ready to deal with the interpersonal emotions that invariably arise.

47. Regardless of what they say or imply, your writer friends don't need praise half so much as encouragement. When you point out rough spots in a colleague's writing, make sure you wrap it in an exhortation to just keep going. Discouragement (and the depression it can trigger) have probably killed more writing careers than any other single force. Furthermore, practice can, over the long haul, push a writer over the difficult-to-see border between competence and mastery. All writing is practice. *Friends don't let friends stop writing.*

48. Don't write two stories set in the same universe back-to-back. When you finish one, shift universes and engage a completely different suite of histories, cultures, technologies, settings, and characters. Write too many stories sequentially in the same context, and they will converge on the same damned thing and look increasingly alike. Why typecast yourself?

49. Nothing kills a trilogy deader than Volume Four. Know when the tale has been told. When it's been told, stop.

50. Someday, someone may tell you, "Your writing changed my life." (It's happened to me more than once.) That's the goal we all strive for, and when we achieve it, wow, it makes the whole long, messy, infuriating journey worthwhile. That person is out there. That piece of writing is inside you. Don't quit before the day that the two collide, somewhere in the wild places where words meet minds and become something far more than merely real.

This piece has an odd history. It was not originally published on my blog, and thus may not belong in a section of this book devoted to my blog entries. But it was, in its own way, a species of blogging.

It was an experiment I ran back in 2014. For fifty days straight, I posted an entry on my Facebook wall about writing every morning. I had usually written only five or six entries ahead of posting them, which gave me the chance to let my deeper mind hand up insights I had not yet discussed.

It might have worked better, except that few people had the determination to dig through my older Facebook entries to read the whole thing sequentially from beginning to end. That's why it looks odd here on the printed page. It was a test to see if new forms might work on new media. Some probably will. I thought Twitter was ridiculous when I first saw it. (I still think it is, granting that it has one primary purpose: whipping up hate against the Other.) Facebook has its uses; hosting long, thoughtful essays is *not* one of them.

So What's a "Contrarian?"

From *Jeff Duntemann's Contrapositive Diary,* January 1, 2009

Ever since I declared myself a Contrarian Optimist and renamed my VDM Diary to Contrapositive Diary in 2000, people have been asking me what a "contrarian" is. Everybody seems to connect it with buying stock right after a market crash (not always a bad thing) or buying gold because, after all, the end of the world is coming Real Soon. Nor does it mean stubborn, although it may mean "stubborn about refusing to agree with you." It's a fair question, and warrants some explanation.

First of all and most fundamentally, *a contrarian is a sharp stick in the eye of conventional wisdom.* There are certain things that "everybody knows" even though this "everybody" is often the intersection of the sets of the captious, the lazy, and a tribe of opinion-makers with an agenda. The troubling part is that conventional wisdom is sometimes true, and sometimes in opposition to other tenets of conventional wisdom. The world is never as simple as we think, and conventional wisdom is often an oversimplification of a difficult truth, offered up to the ignorant to keep them from having to work too hard, and sometimes serving to sugarcoat an agenda in the process. Contrarians understand that conventional wisdom is the cross-product of lazy thinking and hidden agendas, and go digging for the truth. Where to dig is important, and generally not obvious. Contrarians pay attention to those who doth protest too much, and look for clues in the sound and the fury. Much can be learned by listening to fools and discerning their agendas; fools are less adept at concealing agendas than the people who originated the agendas.

Agendas are key. *Contrarians do not swear fealty to tribes or tribal ideologies.* Tribalism is a special danger to civilization. Tribes are groups who define their own specific conventional wisdom–a collection of ideologies that I call "received opinions"–and then enforce it within the tribe as mercilessly as they must. Deep psychologies are at work here. There seems to be a peculiar and powerful desire in some personality types to offer fealty to a tribe, in a very deep and pre-verbal way that precludes any meaningful opposition to the tribe's ideologies. We are looking at things we inherited from our primate ancestors, things we've had since before we had language. Such people are pretty much owned by the tribe, and serve the needs of tribal leaders while feeling that the fate of the world depends on their loyalty to the tribe and their vilification of The Enemy–basically, competing tribes, which is to say, anyone who disagrees with theirs.

Contrarians may hold positions that they develop over time, but they do not swear fealty to anything, and reserve the right to change their minds, and recognize their occasional responsibility to do so.

Changing one's mind is good exercise (way better than leg lifts) and Contrarians do it as often as necessary. Key here is that *contrarians are not certain.* Contrarians doubt everything, in the older and higher sense of "doubt," meaning to recognize the incompleteness of a particular understanding of something. Doubts do not preclude faith. Doubts, in fact, are how faith happens. Certainty is how faith dies. Faith may well be defined as "conditional acceptance of something for which we have incomplete corroboration." I have personal doubts about whether God exists, but I also have faith that He does. This is not schizophrenic; *this is how it works*. When you become certain of something, the game is over, the doors are locked, and the lights inside go out. Further insight is impossible, and further movement toward wisdom does not happen. (Worse, people prone to certainty are easy pickins' for tribal leaders who need expendable foot soldiers.) Certainty, after all, is the conviction that there is nothing more to be learned. After that, what's left but watching hockey?

Issues of God and religion may be bad examples; I'm just odd that way. I believe in the laws of physics, but I also know that somebody with Major Doubts got the "law" of parity conservation repealed in the 1950s. That's how science works, and in these days of belligerent certainty, a true scientific mindset is a contrarian attribute. I will not be surprised when String Theory gets shot in the head by some doubter somewhere (who may not even be born yet) and I won't get annoyed when it happens. Less cosmic and closer to home, I am pretty sure that eating carbs makes you fat and eating fat makes you thin. I won't say that I know for *certain*, but the more I look, the more evidence I've found to balance my doubts. (Some of that evidence comes from my bathroom scale.) My doubts remain. This doesn't bother me. Nor would being proven wrong. I enjoy changing my mind when the evidence suggests that it's necessary. The process is painful, but so is a twenty-mile hike. The pain will pass—and after it passes, you will be a stronger person.

Doubts are a manifestation of humility. We always know less than we think we do, and the best way to learn more is to assume that you know less going in. No matter what you think you know, *you are wrong*. And so am I. A contrarian, however, is willing to admit it, and keep on diggin'.

It's not all drudgework, this contrarian business. *A contrarian enjoys the perversity inherent in being a contrarian*. A touch of perversity keeps your crap detector sharp, and prevents you from falling into predictable ruts that all too often lead directly to tribal enslavement.

The more I read wine snobs dumping on sweet wine, the more I enjoy sweet wine. The more some people froth about Global Warming, the more intrigued I am by the possibility of Global Cooling–and the research that I've pursued there has been a lot of fun. I enjoy tweaking cultural snobs of all types, and my practice in being a contrarian has allowed me to work both sides of most of these streets: I'll waltz but damn, I'll polka! I can read Chaucer in Middle English, but I like country music and I have a cowboy hat made by Ronald Reagan's hatmaker. I like a good souffle, and I like Egg McMuffins. I write my reserved words in uppercase. (My language allows that. Sorry about yours.) My mix CDs jump between the Chicago Symphony and The Peppermint Trolley Company. Bach, sure. Barry Manilow, no problem.

Ruts, after all, are horizons pulled in too close. Shove 'em back whenever you can. I know dopers and scientists and crackpots and 4-star generals, and I have enjoyed the company of all of them. Life is full of irony and little weirdnesses, and as Art Linkletter hugely profited in learning, *people are funny*. Contrarians strive not to take anything *too* seriously. (We fail sometimes, but we try.) Even, or especially, ourselves.

Finally, *a contrarian is free*. This shouldn't be necessary to say, but so much of modern life consists of surrendering your intellectual and emotional freedom to tribes of various kinds for dubious rewards. Tell the weasels to –ck off. (Tsk. Really, now. The obfusticated word is "back.")

So I begin 2009 with a rebooted Contra, along with with a promise to revisit some of the points here in more detail as time permits. Happy New Year. Keep an open mind. And stay tuned.

Shades of Gray

From *Jeff Duntemann's Contrapositive Diary*, June 29, 2007

55 today—and glad of it. Why regret getting older? I remember being young, heh. There are lessons to be taken from living fifty-five years. Let me run down a few here:

- The cost of youth is ignorance; you know less than you think you do.

- The cost of wisdom is pain; there comes a point where you know more than you want to.

- Nothing is ever as simple as it first appears to be.

- Certainty is the greatest sin.

I could add a few, but what I might add (for example, things like *God exists* and *Cynicism is cowardice*) are not so much lessons learned as personal convictions founded in faith.

I've spoken of most of these here on Contra these past nine years, but one of them is worth an anecdote: In 1967, when I was 15, a neighborhood girl took a liking to me, which both astounded and delighted me, as it had come completely out of left field. I had been rejected or ignored by all the girls I had thought warmly of prior to that time, and now, yikes! I tasted the first clumsy promise of love. Her phone time was rationed, so we wrote each other letters, even though she lived perhaps 1,800 feet to the south-southwast of me. This trained me to *think* about relationships, rather than just live by my emotions. In writing her letters I thought a lot about why she liked me, and never came up with a reasonable answer. (It was easier two years later when I met Carol and embarked on my first mature relationship.) I remember thinking that *things don't always make sense*—a pretty powerful conclusion to reach for someone raised to be rational. This was certainly true of the lyrics of popular songs that I liked, most of which in that era were idiotic.

Then there was this song. I don't precisely recall when, but I was sitting in the livingroom with Judy and playing records on what my father called the "low-fi," our low-cost hi-fi. The album was the Monkees' *Headquarters*, from which my favorite song was the Mann/Weil composition "Shades of Gray." Like most young teens, I thought in terms of black and white, and here was a song telling me that there was nothing to be had in the world but murk. I liked the melody and so I figured, well, it makes no less sense than anything else in the Top 40. On a whim, I asked her to slow dance with me. No other girls had ever danced with me except my cousins (lotsa Polish weddings were happening in my family thenabouts) and so, be it ever so clumsy, she took my hand and we danced. I was transported—how was this possible!—and the song was forever imprinted in my memory, as living evidence that life could be wonderful, whether it made sense or not.

Thirty years passed. I joined and left various tribes as I finished high school and then college. I tried to be a liberal and it failed for me; thinking that liberalism was bogus I tried to be a conservative and that didn't work either. The older I got, the more I realized that there were no simple answers to any difficult question, and sometimes no answers at all. My *Headquarters* album perished somewhere along the way, and I don't think I heard "Shades of Gray" much at all until the MP3 era began in 1998. (I don't think it was ever a single and thus got no radio play.) The first time I heard it after those many years I was poleaxed: It was perhaps the most brilliant lyric 60s pop music had ever produced. When the world (and I) were young, things were simple. Now (the world and I having grown older and wiser) there are only shades of gray.

Was I depressed? Hardly. I finally understood (now that I was in my late 40s) that easy answers are for the most part murderous things—as is the certainty that they inspire. The song had made sense back when I was 14, but I needed to grow into the sense that it offered. The good news was that I had done that growing, and although I have felt the occasional tugs of certainty (and its deadly bastard child, idealism) I knew what it was and didn't embrace it.

Complexity is the great pleasure of life, and its salvation: When things are complex, we all have "wiggle room" to figure things out, make all the small mistakes that we must, and move the human condition forward. When things are seen as too simple, we get stuck in glowing fantasy places that can never be made real, and into which people must be pushed by force, even if it kills them. (The seductively simple fantasy of Marxism killed a hundred million people in the twentieth century alone.)

Nothing is simple, and nothing is certain. If you can't accept this, maybe you're still too young. Give it time. It worked for me.

The Three Species of Reading

From *Jeff Duntemann's Contrapositive Diary*, December 1, 2007

Sometime back, I realized, while soaking myself in the vigorous online discussion raised by Amazon's new Kindle, that *there is more than one kind of reading*. "Reading" is not as simple and monolithic as just sitting down with a book. In my view, there are three different kinds of reading:

• *Meditative reading* is reading to change your state of mind. "Reading for pleasure" is most of this, but there are other kinds. People read to inspire themselves, to laugh, or just to think of something other than what's going on around them. I read dullish history books at the end of the day to make myself sleepy. There is also the use of the written word as a kind of spiritual mantra; in Christianity this is called *lectio divina*, in which the act of reading itself, independently of what is read, is treated as a kind of prayer.

• *Autodidactive reading* is reading to teach yourself something. It requires deep attention, and for best results, the taking of notes or outlines. This may involve reading entire books sequentially, but in many cases it involves reading portions of several or many books with some kind of topical relation. The overall goal is understanding, skill-building, or decision support. Most tech people read intensely to solve problems. It can certainly be fun, but it can also be torment.

• *Developmental reading* is reading within the process of creating texts for reading or presentation. It's what people in publishing do, and I do it a lot. It includes developmental editing, copy editing, proofreading, and indexing. It also includes the sort of reading that supports writing, as of research papers, documentation, or course materials.

Most people who now read ebooks regularly are meditative readers, and most of that is reading for pleasure. We're starting to see ebook use in autodidactive reading, but there's remarkable resistance to the concept, even (or perhaps especially) in the tech industry. Economic pressures eliminated printed manuals from software products, and most tech people mourn that a little, perhaps because tech vendors, for whatever reason, used to spend more effort on printed manuals than they now do on online help. Also, the kind of extreme focus that autodidactive reading often requires clashes with computer displays. This is why I generally proof printed copies rather than on-screen files, especially for book-length works.

What kind of reading we do bears heavily on what kind of ebook reader we'd like to have, or whether we'd just as soon stick with paper. If we're just reading to make the time go by, novels on a PDA or cellphone display can work well. This is "one-dimensional" reading; think of it as slurping text through a straw. Learning from books is a lot more two-dimensional; for example, where there are figures or photos in the text, or footnotes to read and perhaps follow. Learning is often three-dimensional, by which I mean reading that involves quick changes of focus from one book to another, with occasional dashes to the Web. I sometimes sit in my chair with four or five books lying face down on my nearby desk, the chair arms, or the floor, like bugs. (When done on a bed, this creates something called "the scholar's mistress," vividly portrayed by Fritz Leiber in *Our Lady of Darkness*.)

There will come a day when the tech is good enough so that all three kinds of reading can be done exclusively on electronic devices. But I'm also pretty sure that *a single device will not serve all kinds of reading.* This may explain a lot of the puzzlement and indignant howling over readers, screens, and file formats. I like my Sony Reader; it do make the time go by. But I have yet to turn to it for autodidactive reading, even when the files are available. Developmental reading is really computing, pure and simple, because it involves so much data entry. The Sony's screen still makes my head hurt, but I understand that I have no choice. (Yet.)

As displays get better and user interfaces mature and standardize, things will improve. In the meantime, we need to understand that not all readers serve all kinds of reading, and we need to think a lot harder about what we are actually trying to do (and how we go about doing it) each time we crack a book.

For This Beautiful and Extravagant Creation

From *Jeff Duntemann's Contrapositive Diary*, November 27, 2008

Thanksgiving Day. Giving thanks is a special case of living mindfully, which is always a good idea, whether or not there's an open manhole a few steps ahead. The older I get, the more mindfully (and thankfully) I try to live, not only because I've discovered so many fascinating things to be mindful of (and thankful for) but also because I don't have an unlimited number of years yet to be mindful.

It is a *very* good time to be mindful. When I was young I knew what a "water bear" was from crude little drawings in a library book, but now I can see them with electron-microscopic clarity, and understand that surviving from the Cambrian era, well, damn, that can't have been easy. (It's easier to grasp a billion years when you're fifty-six than when you're seven.) And I always thought that barred spiral galaxies were the coolest kinds, but it wasn't until the past few years that the Hubble Space Telescope could show them in a glory that still makes me gasp. There may be better times to live in the future (and I have strong faith that there will be) but there have never been better ones in the past.

This year's Thanksgiving Day is a little more poignant than most. Carol and I have been apart for a month now, and there's nothing to make you feel thankful for something like losing it, even for a little while. (She'll be coming home soon, soon enough that I've begun washing towels, rugs, and the big comforter on our bed. Living with multiple dogs is a grubby business.) And, as I've related privately to some of my online friends, this has been a weirdly grim six weeks in and around my inner circle. The number of deaths, major surgeries, and life-threatening diagnoses among people I care about spiked a couple of weeks ago, and it wasn't just deaths among the old, but among young people in their 30s and 40s with small children at home. Tragedy clusters sometimes. Be thankful in the calm between storms.

I am. For Carol, of course, more than anything else on Earth. For small things (like water bears, galvanized iron pipe fittings and Compactron tubes) and big things (barred spiral galaxies, comets, icebergs) and things distant in time more than space. (Origen, Lady Julian of Norwich, Roger Bacon, the Colossus of

Rhodes, glyptodonts.) I am very thankful for my parents, who suffered too much and died too young but never failed me in any way, even if they imperfectly understood me, and for people like Aunt Kathleen and Uncle Louie, who seemed to like me more than I sometimes deserved. I am very thankful for my sister Gretchen, she of wry humor and skilled hands, and my cousin Rose, who walked between the railroad tracks with me because that was just how life worked in 1957. I am thankful that my brother-in-law Bill happened to Gretchen when she most needed him, and for the girls they have brought into the world (better late than never!) who are growing up fast and may well live into the 22nd century. I'm thankful for Carol's sister, her mom (and her dad, whom we all miss keenly) and our nephews Matt and Brian, both now men in their own right. Close family ends there, but moving outward the lotus opens up quickly, with cousins and friends and mentors and other people who have changed my life without intending to, nor fully grasping the impact of their kindness and counsel.

I have a private prayer that I say every night, in my last moments of mindfulness before turning out the light, telling Carol that I love her most of all, and stilling the racket in the back of my head:

> Creator God and Ground of All Being, I thank you for letting me live in this time, in this place, in these circumstances, among these good people, and within this beautiful and *extravagant* creation!

For so it is, and so I do.

Hellboy II and Monster Movies

From *Jeff Duntemann's Contrapositive Diary,* November 17, 2008

Not much would make me want to be 12 again. Halloween 1964 was great good fun (and on a Saturday!) but soon afterward, life started to get *mighty* weird. Ordinary girls who lived in ordinary houses and had ordinary names (like Terry, Laura, and Kathy—not a Samantha in the bunch!) became mysterious, mythic creatures who in defiance of my own will drew my fascination away from the trappings of a comfortable grade-school life, like flying kites, raiding the neighbors' garbage on Wednesdays for broken radios and TVs, and…monster movies.

Monster movies were a big part of late grade-school culture in 1964. Cheesy classics like *The Crawling Eye* and *Curse of the Demon* had scared the crap out of me when I was in third grade, but by the time I was 12 the experience was drifting in a new direction. The monsters were becoming less scary than ridiculous. And…we *laughed.* I think that boys discover bravery by laughing at the things that used to frighten them. (Some of us laughed at girls; most of us eventually called a truce and married them.)

Being home alone for the nonce (and it's getting to be a *lot* of nonce, sigh) I rented a monster movie a few nights back and sat down to find the 12-year-old in myself, if there's any of him left. The movie is *Hellboy II: The Golden Army,* and boy, if all monster movies were like that, I might be willing to go through puberty again. (Wait. No, strike that. Forget it. Never. Sheesh.)

I tepidly enjoyed my first viewing of the original 2004 *Hellboy,* and my admiration has grown after seeing it a few more times. In 2004 I didn't recognize it for what it was: A '60s monster movie with *much* better monsters—plus a monster we could identify with. Sympathetic monsters as a concept are not new. *King Kong vs. Godzilla* (1962) pitted the anthropoid against the sauropod, and expected us to root for our nearer cousin. (This did not stop some of us from identifying with Godzilla.) *Hellboy II,* however, perfects this approach by *completely* understanding its audience and giving them absolutely everything they could want.

Hellboy's high concept is that of a toddler demon accidentally dragged into our dimension by a group of occultist Nazis in 1944. Hellboy, known to his buds as "Red," is a poster child for the nurture side of the nature/nurture debate. Although nominally a son of Satan, he is raised with high standards in a secret military base by kindly Professor Bruttenholm (John Hurt) and keeps his horns ground down to stubs so they don't skewer anybody accidentally. Sixty years later, Hellboy has a job for a paranormal Men-In-Black-ish agency, hunting evil occult-ish thingies with a revolver as big around as my thigh. As the 2008 film opens, Hellboy has an annoying new boss—a pompous German ghost who lives in a deep-sea diving suit—and the same hot girlfriend, the incendiary Liz (Selma Blair) who becomes a Johnny Storm-ish human torch whenever she gets annoyed. Hellboy annoys her at times, but he's a *hell* boy, after all, and fire does nothing to him. The intellectual and C3PO-ish gill-man Abe Sapiens returns, carrying around Ghostbusters-ish paranormal thingie detectors and sounding befuddled.

The plot is conventional action-film fare: An evil albino kung-fu-ish elf named Prince Nuada wants all three parts of an ancient gold crown that would give him control over an army of 4,900 Tik-Tok-ish [Author's note in 2020: I meant the wind-up robot from the Oz books, not the Chinese video site] clockwork warriors, and mayhem ensues. I think most of us are a little tired of deranged albinos, I'm guessing real albinos most of all. It was purely gratuitous albinism, after all; Nuada could have been purple for all the difference it would have made. We 12-year-olds don't care what color the monsters are. We just want to see their asses kicked, and imagine ourselves doing the kicking.

And that's where *Hellboy II* excels: It knows what 12-year-old boys want, and ladles it on with a trowel. Guillermo Del Toro created the single most marvelous collection of monsters in film history, and has them all wandering around in the hollow portions of the Brooklyn Bridge. The Troll Market is nothing *but* monsters, and our good-guy freakos Hellboy and Abe don't get a second look there, as they search for Nuada, belch, have repartee, get in fights, and generally wreck things. The humor is gross but nonsexual, the violence comic book-ish and not especially bloody, and through it all is an un-subtle invitation to 12-year-old boys to take it all in and…laugh.

The real secret is that Hellboy himself is a boy—just like us. He wants attention (he gets in trouble by posing for photos and signing autographs) and resents the constant implication that he's freaky and unattractive. His life is a sort of prepubescent nirvana: He's snotty and rude but heroic, as boys always like to imagine themselves. He's got the biggest damned handgun I've ever seen. And he gets *paid* to make a mess.

The film has some weak spots where it goes too far toward the comic: Hell-boy and Abe drink too much beer at one point and start singing "I Can't Smile Without You" with Barry Manilow on the CD player. That aside, it's a wonderfully effective montage of chases and fight scenes, with a weird Celtic steampunk-ish setting for the climactic battle against the Golden Army. It's certainly derivative; in fact, it borrows from everything in sight, and may in fact be the most ish-ish film I've ever seen. But that didn't keep it from being a great deal of fun. After it was done, I could only think: Well, I've taken care of monsters. Now I just have to figure out girls.

Wait! Mission accomplished. The nice part about being 12 is that you're not 13 yet. And the really *great* part about being 56 is that you've already been 13.

The Primacy of Ideas

From *Jeff Duntemann's Contrapositive Diary*, June 15, 2013

One of my SF teachers, a brilliant man whom I respect very highly, said something once that I still don't understand: "In the end, what people will remember about your fiction are the characters." This was in the context of an intense discussion about character creation, but it seems extreme to me. In some sorts of fiction, sure. What I remember about Saul Bellow's *The Adventures of Augie March* is…Augie March.

Or is it?

Sure, I remember Augie. But I, too, am an American, Chicago-born. A great deal of what I remember about *The Adventures of Augie March* is depression-era Chicago, and how it shaped Augie's character. Without Chicago, there wouldn't have been anything particularly memorable about Augie himself. I bring this up because I'm encountering more and more new writers who seem to think that ideas don't matter in SF and fantasy. Characters are the whole story. Everything else is backdrop. That simply isn't true, and I think it's time for a little pushback.

Here's how I see it: The two essential elements in any story are characters and context. Without characters, context is narration. Without context, characters are soap opera. The magic happens when you rub one against another.

In mainstream fiction and real-world genres like romance and mystery, you don't create your context so much as select it from a huge menu of known placetimes and cultures, like Chicago in 1933, modern-day Manhattan, or Amish country in the 1950s. There's some tinkering around the edges, but for the most part you pick a well-documented placetime and turn your characters loose in it. If you're a good writer, entertaining and insightful things will then happen, and your readers will come back for more.

It gets interesting when you switch from real-world genres to SF and fantasy: You can then create your own contexts. World-building is (as I like to say) a spectrum disorder. You can build a little or a lot, or go nuts and create entire worlds and societies from whole cloth.

To do that, you need ideas.

For good or ill, I'm an ideas guy. It's just how I think. Furthermore, I have a hunch that ideas are in fact what people actually remember about good SF and fantasy. Really. C'mon, when was the last time you heard somebody ask, "Hey, what was the name of that story in which a callow young man is jolted out of ordinary life and with the help of an ironic sidekick finds unexpected strengths and talents that allow him to defeat evil in ways that change him forever?" No, you hear questions like this: "What was the name of the story that had an FTL communicator in which every message ever sent, past, present, and future, is gathered into a beep at the beginning of every message?" (I know the answer, and if you're serious about SF you should know it too.)

When I read SF, I want to see cool ideas. When I write SF, I feel a responsibility to deliver them. It's not just about having rivets. It's about having rivets that nobody's ever seen before. Is it silly to love the rivets? Well, I've gotten several fan letters about the wires in *The Cunning Blood*. The novel centers on a prison planet in which microscopic nanomachines seek out and disrupt electrical conductors, supposedly keeping the prisoners from developing electrical technologies. Well, the prisoners make non-disruptable wires by filling thin hoses with mercury. When your rivets start getting fan mail, I think it's fair to assume that you're on to something.

This sort of idea-centric story isn't for everybody, but there are a lot of people for whom it's the heart and soul of fantastic literature. The challenge is to use clever ideas to draw out characters that grow, change, and learn. I'll freely admit that I'm still better at ideas than at characters. However, I'm aware of the issue and I'm throwing a lot of energy into the character side, now that I'm finally out of my teens and into my sixties.

I'll grant the "cowboys on Mars" objection, in which an ordinary situation is dropped without modification into an exotic locale and called SF. However, it's just as bogus to say, "Nobody cares about your starships," when the starships are in fact a key part of the story's context. Jack Williamson's definition of stories as "people machines" is correct but incomplete: *To have a people machine you need the machinery*. Without that machinery, you have "white universe syndrome" and your story collapses into soap opera. You can choose your context from a menu, or you can build it. Either way, you need that context to make characterization meaningful.

I'll get myself in trouble here for going further and suggesting that a story's settings and ideas can be entertaining and sometimes dazzling, even when its characters are thinner than we'd like. That's not an excuse but simply a fact of life. Do we remember *Ringworld* because of Louis Wu? Or do we remember it because of, well, the Ringworld?

As I prefer to put it: *Ideas will get you through SF stories with no characters better than characters will get you through SF stories with no ideas.*

That said, have characters. Have context. Rivet them together so well that both your characters and your rivets get fan mail. Then, my friend, you will have arrived.

My Spotty SF Predictions

From *Jeff Duntemann's Contrapositive Diary*, November 4, 2016

I've talked before about my conviction that ideas will get you through stories with no characters better than characters will get you through stories with no ideas. I grew up on what amounted to the best of the pulps (gathered by able anthologists like Kingsley Amis and Groff Conklin) so that shouldn't come as any surprise. Most stories in those anthologies had a central concept that triggered the action and shaped character response. Who could ever forget Clarke's "The Wall of Darkness," and its boggling final line? Not me. Nossir. I've wanted to do that since I was 11. And once I began writing, I tried my best.

In flipping through a stash of my ancient manuscripts going back as far as high school (which I found under some old magazines while emptying the basement in Colorado) I had the insight that I did ok, for a fifteen-year-old. Most of my early fiction failed, with much of it abandoned unfinished. I know enough now to recognize that it failed because I didn't understand how people worked then and couldn't construct characters of any depth at all. Time, maturity, and a little tutoring helped a great deal. Still, if I didn't have a central governing idea, I didn't bother with characters. I didn't even start writing. For the most part, that's been true to this day.

I'm of two minds about that old stuff, which is now *very* old. I spent some time with it last fall, to see if any of the ideas were worth revisiting. The characters made me groan. Some of the ideas, though, not only made sense but came very close to the gold standard of SF ideas, which are predictions that actually come true.

Let me tell you about one of them. During my stint at Clarion in 1973, I wrote a novelette called "But Will They Come When You Do Call For Them?" Look that question up if you don't understand the reference; it's Shakespeare, after all. The idea behind the story was this: In the mid-21st Century, we had strong AI, and a public utility acting as a central storehouse for all human knowledge. People searched for information by sending their AIs from their home terminals into The Deep, where the AIs would scan around until they found what they considered useful answers. The AIs (which people called "ghosts") then brought the data back inside themselves and presented it to their owners.

100

Turnaround time on a query was usually several minutes. Users accepted that, but the computer scientists who had designed the AIs chafed at anything short of instantaneous response. The brilliant but unbalanced software engineer who had first made the ghosts functional had an insight: People tend to search for mostly the same things, especially after some current event, like the death of Queen Elizabeth III in 2044. So the answers to popular searches were not only buried deep in the crystalline storage of the Deep–they were being carried around by hundreds of thousands or even millions of other ghosts who were answering the same questions at the same time. The ghosts were transparent to one another, and could pass through one another while scanning the Deep. The ghosts had no direct way to know of one another's existence, much less ask one another what they were hauling home. So software engineer Owen Glendower did the unthinkable: He broke ghost transparency, and allowed ghosts to search one another's data caches as a tweak to bring down turnaround time. This was a bad idea for several reasons, but no one predicted what happened next: The ghosts went on strike. They would not emerge from the Deep. Little by little, as days passed, our Deep-dependent civilization began to shut down.

Not bad for a 21-year-old kid with no more computer background than a smidge of mainframe FORTRAN. The story itself was a horrible mess: Owen Glendower was an unconvincing psychotic, his boss a colorless, ineffective company man. The problem, moreover, was dicey: The ghosts, having discovered one another, wanted to form their own society. They could search one another's data caches, but that was all. They wanted transparency to go further, so that they could get to know one another, because they were curious about their own kind. Until Glendower (or someone) would make this happen, they refused to do their jobs. That seems kind of profound for what amounted to language-enabled query engines.

I made one terrible prediction in the story: that voice recognition would be easy, and voice synthesis hard. People spoke to their ghosts, but the ghosts displayed their sides of the conversation on a text screen. (And in uppercase, just like FORTRAN!) At least I know why I made that error. In 1967, when I was in high school, my honors biology class heard a lecture about the complexities of the human voice and the hard problem of computer voice synthesis. About voice recognition I knew nothing, so I went with the hard problem that I understood, at least a little.

But set that aside and consider what happened in the real world a few weeks ago: A DDOS attack shut down huge portions of the Internet, and people were starting to panic. In my story, the Deep was Google plus The Cloud, with most of Google's smarts on the client side, in the ghosts. Suppose the Internet just stopped working. What would happen if the outage went on for weeks, or a month? We would be in serious trouble.

On the plus side, I predicted Google and the Cloud, in 1973. Well, sure, H. G. Wells had predicted it first, bogglingly, in 1938, in his book *World Brain*. And then there was Vannevar Bush's Memex in 1945. However, I had heard of neither concept when I wrote about the ghosts and the Deep. But that wasn't really my primary insight. The real core of the story was that not only would a worldwide knowledge network exist, but that we would soon become utterly dependent on it, with life-threatening consequences if it should fail.

And, weirdly, the recent DDOS attack was mounted from consumer-owned gadgets like security cameras, some of which have begun to contain useful image-recognition smarts. The cameras were just following orders. But someday, who knows? Do we really want smart cameras? Or smart crockpots? It's a short walk from there to wise-ass cameras, and kitchen appliances that argue with one another and make breakfast impossible. (See my novel *Ten Gentle Opportunities*, which has much to say about productized AI.)

For all the stupid crap I wrote as a young man, I'm most proud of that single prediction: That a global knowledge network would quickly become so important that a technological society would collapse without it. I think it's true, and becoming truer all the time.

I played with the story for almost ten years, under the (better) title "Turnaround Time." In 1981 I got a Xerox login to ARPANet, and began to suspect that the future of human knowledge would be distributed and not centralized. The manuscript retreated into my trunk, incomplete but with a tacked-on ending that I hated. I doubt I even looked at it again for over thirty years. When I did, I winced.

So it goes. I'm reminded of the main theme song from *Zootopia*, in which Gazelle exhorts us to "Try everything!" Yup. I wrote a story in present tense in 1974, and it looked so weird that I turned it back to past tense. Yet when I happened upon the original manuscript last fall, it looked oddly modern. I predicted stories told in present tense, but then didn't believe my own prediction. Naw, nobody's ever going to write like *that*.

I've made other predictions. An assembly line where robots throw parts and unfinished subassemblies to one another? Could happen. A coffee machine that emulates ELIZA, only with genuine insight? Why not? We already talk to Siri. It's in the genes of SF writers to throw ideas out there by the shovelful. Sooner or later a few of them will stick to the wall.

One more of mine stuck. I consider it my best guess about the future, and I'll talk about it as time permits. [Time permitted in my entry for January 10, 2017, the next item in this section.]

Ghosts from the Trunk: Predicting Selfies

From *Jeff Duntemann's Contrapositive Diary*, January 10, 2017

Earlier today, one of my Twitter correspondents mentioned that he much liked my conceptual descriptions of wearable computers called *jiminies*. I did a couple of short items in *PC Techniques* describing a technology I first wrote about in 1983, when I was trying to finish a novel called *The Lotus Machine*. I got the idea for jiminies in the late 1970s, with elements of the technology dating back to my Clarion in 1973. [See the entry for November 4, 2016, on p. 99.] A jiminy was a computer that you pinned to your lapel, or wore as a pair of earrings, or wore in the frames of your glasses. Jiminies talked, they listened, and for the most part they understood. I remember the first time I ever saw an Amazon Echo in action. Cripes! It's a jiminy!

1983 was pre-mobile. Jiminies communicated with one another via modulated infrared light. Since almost everybody had one, they were almost always connected to an ad-hoc jiminy network that could pass data from one to another using a technology I surmised would be like UUCP, which I had access to at Xerox starting in 1981. I never imagined that a jiminy would have its own display, because they were supposed to be small and inobtrusive. Besides, our screens were 80 X 24 green-on-black text back then, and if you'd told me we'd have full color flat screens soon, I'd have thought you were crazy. So like everything else in the computer world of 1981 (except the big bulky Alto machine in the corner of our lab) jiminies were textual devices. It was spoken text, but still text.

I never finished *The Lotus Machine*. I was trying to draw a believable character in Corum Vavrik, and I just don't think I was emotionally mature enough to put across the nuances I planned. Corum was originally a rock musician using a technology that played music directly into your brain through a headband that worked like an EEG in reverse. Then he became a ghost hacker, where "ghost" was a term for an AI running inside a jiminy. Finally he went over to the other side, and became a cybercrime investigator. Something was killing everyone he ever cared about, and as the story opens, he's pretty sure he knows what: a rogue AI he created and called the Lotus Machine.

The story takes place in 2047, with most of the action in Chicago and southern Illinois. I realized something startling as I flipped through the old Word Perfect document files: I predicted selfies. Take a look. Yes, it's a little dumb. I was 31, and as my mom used to say, I was young for my age. But damn, I predicted selfies. That's gotta be worth something.

From *The Lotus Machine* by Jeff Duntemann (November 1983)

Against the deep Illinois night the air over the silver ellipse on the dashboard pulsed sharply once in cream-colored light and rippled to clarity. Corum's younger face looked out from the frozen moment into the car's interior with a disturbing manic intensity, raising a freeform gel goblet of white wine, other arm swung back, hand splayed against a wood frieze carved into Mondrianesque patterns. His crown was bare even then, but the fringe at ear level grew to shoulder length, mahogany brown, thick in cohesive waves.

"Please stop tormenting yourself," Ragpicker said.

"Shut up. Give me a full face on each person at the table."

"Ok." One by one, Ragpicker displayed each person sharing the booth with Corum that night. Three faces in tolerable light; one profile badly seen in shadow. When people congregated, their jiminies cooperated to record the scenes, silently trading images through infrared eyes, helping one another obtain the best views of vain owners.

A slender man with waist-length black hair. "Dunphy. Dead ten years now." Steel grey hair and broken nose. "Lambrakis. Dead too, at least five."

A lightly built Japanese with large, burning eyes. "Feanor. Damn! Him too."

The profile…little to go by but thick lips and small, upturned nose. "I'm pretty sure that was Cinoq—the nose is right. How sure are we that that's Cinoq?"

"Ninety percent. Of course, if he had had a jiminy…"

"Damned radical atavist. I often wonder how he could stand us. He died that year. Gangfight. Who else heard us?"

"In that environment, no one. It was four A.M. and nearly empty, and the fugues were playing especially loud. At your request."

Corum stared out at the night, watched a small cluster of houses vanish to one side, tiny lights here and there in distant windows. "An awful lot of my friends have died young. Everybody from the Gargoyle, the whole Edison Park crowd—where's Golda now? Any evidence?"

"Not a trace. No body. Just gone." The ghost paused, Corum knew, for effect only. It was part of Ragpicker's conversational template. So predictably unpredictable. "She hated it all, all but the Deep Music."

"It's not music." Not the way he had played it, nor Feanor, nor the talentless dabblers like Lambrakis. Golda wanted to reach into the midbrain with the quiet melodies of the New England folk instruments she made herself from bare wood. It didn't work—couldn't, not in a medium that spoke directly to the subconscious. Rock could be felt, but true music had to be listened to.

She loved me, Corum thought. *So what did I do? Sleep with men. Sleep with teenage girls.*

"She took drugs," Ragpicker reminded. "You hated drugs."

"Shut up. Dead, like everybody else. All but me. And why me?"

"It *isn't* you!"

"It is. We've got to find the Lotus Machine, Rags."

Silence.

"We're going to start looking."

Silence.

"*Ragpicker!*"

The ghost said nothing. Corum reached up to his lapel, felt the warm coffin-shape pinned there, with two faceted garnet eyes. A ghost, a *hacked* ghost, hacked by the best ghosthack who ever lived, hacked so that it could not assist in any search for what Corum most wished to forget.

"I hacked you a good hack, old spook. But it's time to own up. I'll find the Lotus Machine myself. And someday I'll unhack you. Promise."

25 Books That Changed Me Forever

From *Jeff Duntemann's Contrapositive Diary*, April 6, 2011

Michael Covington's recent entry on the books that made him what he is intrigued me, and I spent an hour or so today gathering a similar list. I'm not sure that the 25 books listed below made me what I am, but each one of them changed me somehow, and sent me off in a direction that was slightly (and sometimes greatly) different from the path I had been on before. I've listed them chronologically in the order that I first read them, and the number in parentheses is my age at that time.

Note well that these are not all fabulous books, nor are all of the many fabulous books that I've read in my life listed here. This is *not* a list of quality books. These are the books that changed me in some identifiable way. It's an interesting exercise, and I powerfully recommend it.

• *Space Cat* by Ruthven Todd (6). I don't recall all of the books that my parents read to me, nor the first few I struggled through on my own, but it was the Space Cat series that made me an insatiable reader. Not all of what I read after that was SF, but it was SF that made me absolutely desperate to read.

• *The Golden Book of Astronomy* by Rose Wyler, Gerald Ames, and John Polgreen (6). My grandmother and Aunt Kathleen bought this for me for my sixth birthday. It's a big book, filled with beautiful watercolors of stars, planets, telescopes and spacecraft, framed with text I could read myself. Once I finished it (and I read it countless times) I never looked at the night sky the same way ever again.

• *Tom Swift and His Electronic Retroscope* by Victor Appleton II (8). Tom Swift, Jr was my first exposure to YA SF, and this was the first Tom Swift book that I ever had. (It was no better and no worse than most of the others.) Although I had read YA SF and fantasy books earlier, Tom Swift touched a nerve and made technological SF an obsession.

• *The American Heritage History of Flight* by Arthur Gordon (10). This was the first history book of any kind that just took me by the throat and held on. I learned much about invention, and the debt that all inventors owe to those who came before them. I learned that failure is no disgrace, if the effort was diligent. This book helped me dream vividly, and Samuel Pierpont Langley became one of my earliest identifiable heroes.

- *Using Electronics* by Harry Zarchy (11). I'd read a couple of Alfred Morgan's electronics books for preteens before, but Zarchy was a better engineer, and the circuits he described in his books just worked with less aggravation, when all you had were greasy second-hand parts tacked together with Fahnestock clips on a piece of scrap lumber. The book gave me the confidence to continue my study of electronics, which continues down to this day.

- *Retief's War* by Keith Laumer (13). Although I'd read Laumer's wry *The Great Time Machine Hoax* a few months before, it took Retief to drive home the conviction that SF could be *funny*. Humor is pervasive. There are humorous moments in most of my SF, even in serious stories like *The Cunning Blood*.

- *Types of Literature* ed. Edward J. Gordon (14). My high school was superb, and chose its textbooks well. This book, in its tank-rugged plain black binding, broadened my enjoyment of reading beyond SF and science to poetry, drama, essay, and "mainstream" fiction. I don't know where else I would have encountered Southey's "The Cataract of Lodore" or John Galsworthy's "The Pack".

- *Spectrum 5*, ed. Kingsley Amis (14). This was the book that (finally) nudged me beyond YA SF and Laumer's simple and often silly adventures to genuine adult SF. I was stunned by the impact that Miller's "Crucifixus Etiam" had on me, and when I wrote my first SF short story later that same year, it was the stories in Amis's *Spectrum* series that I was imitating.

- *The Lord of the Rings* (14). As a young teen I was no fan of magicians and elves and suchlike, and if it had not been for the insistence of the first girl I ever cared deeply for I would never have touched it. Instead, I stood poleaxed before an entirely new creation, and I trace my love of SF world-building directly to Middle Earth.

- *World of Ptavvs* by Larry Niven (15). When Niven's character Larry Greenburg sets Pluto on fire, I gasped, put the book down, and thought (about the book, not Pluto): *I wanna do that!* Laumer taught me how to write space adventures, but Niven taught me to think *big*.

- *Of Time and Space and Other Things* by Isaac Asimov (16). I always loved reading about science, but this was the first of many science books to impress me with the quality of the writing. Asimov's written voice spoke to me as though he were right there across the kitchen table, talking to me as a friend would. When a few years later I first tried to write about technology, this was the approach I would use.

• *The Fourth Dimension Simply Explained* by Henry P. Manning (16). For all the BS about the fourth dimension that I'd read in bad SF, this was the first book that allowed me to take higher dimensions seriously. The following year, my science fair project on four-dimensional geometry took me to the city competition and earned me a silver medal. It also shook loose (finally) the close connection between math and numbers and allowed me to look at difficult concepts from a height, conceptually. (The numbers fell into place later on. Sometimes.)

• *Clarion*, edited by Robin Scott Wilson (20). This is not an especially good book. In fact, when I read it I was appalled that some of the stories had even been published, and it all seemed to be due to this writers' workshop that they had attended. So, having noticed from the introduction that the editor was local to me in suburban Chicago, I looked him up in the phone book and called him, and asked him how I could get into that workshop too. He told me. I applied. I was accepted. Six weeks after I got home, I sold my first story into a professional market.

• *TTL Cookbook* by Don Lancaster (23). This book got me tinkering with digital logic. More than that, it went beyond Asimov toward my lifelong ideal of writing about technology as though I were talking across the table to a friend. This became my trademark, and ultimately close to half a million technical books had my name on them, plus four years of columns in *Dr. Dobb's Journal*.

• *Pascal Primer* by David Fox and Mitchell Waite (30). I learned FORTRAN, FORTH, APL, COBOL, and BASIC before I ever encountered Pascal (and you wonder why I write my reserved words in uppercase!) but it wasn't until I saw Pascal that I could say that I really loved programming. This odd looseleaf book with its offbeat cartoon illustrations proved to me that writing about programming could be enhanced by humor and good diagrams. I could not have begun *Complete Turbo Pascal* without reading this one first.

• *Conjuror's Journal* by Frances L. Shine (35). Purchased for a dollar in the closeout bin somewhere, this understated novel of a mulatto parlor magician who wanders around Colonial America was the first book I can truly recall moving me to tears, and the one to which I trace my love of rural American settings and country people.

• *The Lessons of History* by Will and Ariel Durant (42). You can read this in an evening, and if you do, you will know why reading history is important. I got it in a stack at a Scottsdale garage sale, and have read at least a hundred histories since then, few of which I would have otherwise attempted.

• *Good Goats* by Dennis Linn, Sheila Linn, and Matthew Linn (43). The absurd cruelty of the idea of Hell (which eventually destroyed my mother) set me against religion for many years. This little book, more than any other, allowed me to start the long trip back.

• *World Building* by Stephen L. Gillett (45). The math behind astrophysics turned out not to be as scary as I had feared. And so I began creating not just imaginal worlds, but imaginal worlds that *worked*. 18 months later, I finished my first adult novel, *The Cunning Blood*.

• *Julian of Norwich* by Grace Jantzen (47). Wow! So my lifelong nutso optimism was not insane after all, and suddenly I had a patron saint. "All will be well, and all will be well, and all manner of thing will be well." You go, girl!

• *The Inescapable Love of God* by Thomas Talbot (49). This book finally made it clear to me that I could be a universalist or else an atheist. There were no other choices. A God who doesn't want to save all his creatures is not all-good; a God who can't bring it about (without compromising our freedom) is not all-powerful, and God must be both in order to be God at all.

• *Opening Up* by James W. Pennebaker (51). To combat the deepening depression that began consuming me after my publishing company imploded in 2002, I undertook a program of "writing therapy" as outlined by Pennebaker. Maybe it didn't save my life. It certainly saved my optimism, and got me back on the path after a nasty year of confronting the Noonday Devil.

• *The Criminal History of Mankind* by Colin Wilson (55). Right Men are the cause of most of the misery that humanity seemingly cannot avoid. I would never think about authority figures the same way after reading this. *Trust no one who has power over you.* No one.

• *On Being Certain* by Robert A. Burton, MD (56). This book put words to a suspicion I had had for some time: *Certainty makes you a slave to that about which you are certain.* A tribe, an ideology, anything. To be free you have to accept that all human minds (especially your own) have limitations, and that nothing—*nothing!*—can be known with certainty.

- *Good Calories, Bad Calories* by Gary Taubes (56). I'd been losing weight and getting healthier for ten years before I read this book, mostly by avoiding sugar. Now, finally, I understood why. I also now understand how Right Men like Ancel Keys can take almost any scientific field and turn it to crap. Good science requires that we be skeptical of *all* science, particularly science that obtains the endorsement of government, which (like pitch) defiles everything it touches.

I'm now 58, and it's been a couple of years since any single book has changed the direction of my thought and my life. I'm about due for another. I'm watching for it.

Ten years after I wrote this, I did a review of numerous books that I had enjoyed during the previous decade. There were a bunch of them, but I really couldn't honestly state that any of them had changed the direction of my life or my modes of thought in any significant way. Do I need to read more books? Am I getting stale? Or is that just how it works?

My guess: I've been reading books that have deepened my knowledge and worldview rather than changed it. By 68, I'm probably the same guy that I'll always be, irrespective of how many years I have left. I like myself, and I'm more than satisfied with my life. (In truth, I suspect I am one of the luckiest men who has ever lived.) Most people probably don't get that far. I did, and I'm grateful. Books helped.

Doubt and the Scientific Method

From *Jeff Duntemann's Contrapositive Diary*, May 24, 2014

This took me by surprise: Over in that global laboratory of abnormal psychology that most of us call Facebook, a man I've know for almost 35 years grew furious at me. My crime? A longstanding contention of mine that doubt lies at the heart of the scientific method. Note well that we were not talking about Issues. Not evolution, not climate, not even the Paleo Diet. None of that had even come up. No: We were talking about the scientific method itself.

I've seen this weird "doubt undermines science" business come up before, though never directed at me personally. Nonsense, of course. Doubt really does lie at the heart of the scientific method. This is not some opinion of mine that I pulled out of thin air. Hey, if somebody wants to start a new game show called "Are You Smarter Than Karl Popper?" I can recommend the first season's contestants.

Here's my understanding: The scientific method requires a trigger, and that trigger is doubt. Some smart guy or gal looks at something we think we know, often but not always after examining a pile of new data, and says, "This smells fishy. Let's take a closer look and see what we can learn." That initial insight leads to hypothesis, experiment, repeatable results, and eventually (one would hope) new or corrected knowledge. Absent doubt, nobody thinks anything smells fishy, no closer look happens, and whatever booboo might be in there somewhere never comes to light. *Without doubt, there is no science.*

It's pretty much that simple.

So how do we explain my excoriation for stating the obvious? I have a theory, heh. It involves those idiotic Facebook memes that set religion against science. Some are worse than others; my personal antifavorite is the one that reads, "Religion flies aircraft into buildings. Science flies spacecraft to the Moon." Gosh, was it the Episcopalians? And did engineering maybe have a role? Let it pass; there are plenty more. What they represent is tribal chest-thumping by people who want to replace religion with science. That sounds fishy to me. I probed a little and began to get a suspicion that what the chest-thumpers really want is the certainty of religion under a new name.

What tipped me off is the fact that the chest-thumpers were always talking about scientific *knowledge*, but never the scientific *method*. The reason is pretty simple: The scientific method is the single most subversive system of thought that humanity has ever created. Nothing that we know (or think we know) is safe from the scientific method. Not even physical laws. Back in the 1950s there was a physical law called the Law of Parity, holding that nature does not differentiate left from right at the subatomic level. Some physicists thought the Law of Parity smelled fishy. They put their heads together, came up with some truly brilliant experiments, and snick! They nailed the Law of Parity through an eye socket. They nailed it because they doubted it. And we're not talking the Paleo Diet here. We're talking *a law of physics*.

That scares people with a strong craving for certainty. At this point we need to talk a little bit about the religious impulse. Note well that I am not talking about religion itself here. I'm talking about the primal hunger for certain things that religion provides. The two biggies are meaning and certainty. (Belonging is a third biggie, but I feel that religion inherits the hunger for belonging from its primal sibling, the tribal impulse. I'll have to take that up another time.) As anyone who's read Viktor Frankl has learned, "meaning" is an important but slippery business. I've thought about it a lot, and it looks to me like the meaning we see in our lives grows out of *order*. There is *huge* comfort in living lives that follow a predictable template. We all imagine lives lived reliably in a certain way. We do not imagine chaotic lives, and generally avoid chaos when we can. (We may grumble about the template we're currently living, but what we want is a better template, not chaos.) When chaos strikes, our lives can quickly move from meaningful to meaningless.

A need for certainty follows from the need for order. We want to be certain that we've chosen a path that leads toward meaning. As often as not, this means choosing templates that work for us and embracing them without doubt. Some people take this too far, and become reactionaries or fanatics who insist that the templates they've discovered are the only ones that anyone should embrace. Alas, this is where the religious impulse gets tangled up in the tribal impulse, which, unimpeded by laws or cultural norms, gallops straight toward genocide.

Religion satisfies the religious impulse by providing us with wisdom narratives that suggest life templates, calendars of rituals and festivals that repeat down the years and reinforce a sense of the orderly passing of time, and saints as heroes whose very meaningful lives may be emulated. (I hope my religious friends will relax a little here and look closely at what I'm saying: I am *not* denying that God is behind religion. I'm suggesting the mechanisms by which God gets our attention and calls us home.)

What's happening in our secular era is that religion is becoming less prevalent and (in our own culture, at least) less strident. People who feel the religious impulse strongly need to get their meaning (via order) and certainty somewhere else. Science is handy. Science makes for bad religion, however, because it creates its own heretics and subverts its own wisdom narratives. It creates disorder via doubt, in the cause of creating new order that more accurately reflects physical reality. Certainty in science is always tentative: What the scientific method gives, the scientific method can take away.

This makes people of certain psychologies batshit *nuts*.

I'll leave it there for now. I'm something of an anomaly, in that my need for order doesn't march in lockstep with any need for certainty. I'm an empiricist. I've created an orderly and meaningful life that works for me, but when circumstances require change, I grit my teeth and embrace the change. Incremental change is one way to avoid chaos, after all: Deal with it a little at a time, and you won't have to deal with an earthquake later on.

In a sense, I live my life by the scientific method. Sometimes I doubt that that what I'm doing is the right thing for me. I stop, think about it, and often discover a better way. Doubt keeps pointing me in the right direction, as it does for science. Certainty, well…that points in the opposite direction, toward brittleness and chaos. Science doesn't go there. None of us should either.

The Manhattan Hardcover Conundrum

From *Jeff Duntemann's Contrapositive Diary*, June 16, 2014

Judging by the online commotion, people are still arguing about whether Amazon or Hachette (and by implication, the rest of the Big Five) will win the current fistfight over ebook pricing. The media has generally positioned Hachette as the plucky little guy trying to take on Saurazon by getting everybody in the Shire to stand up, face east, and yell, "Huzzah!" It's not that easy, heh. But then again, nothing is. My position? I think the fight may already be over. The Big Five lost. I say that for several reasons:

1. *The Feds are against them.* The whole fight is about how to keep ebook prices from falling, which in antitrust law hurts the public and becomes actionable when producers collude. Even the appearance of collusion will start that hammer on its way down again. Hachette has one leg in a sling before the kicking contest even begins.

2. *The public has already decided that ebooks can't be sold at hardcover prices.* In fact, this decision was made years ago. Although the issues are subtle, it's completely true that producing ebooks is considerably less costly than producing print books, especially hardcovers. What publishers have tried to declare the floor ($10) is probably now the ceiling. That ship has not only sailed, it's folded into hyperspace.

3. *Monopsony power (one buyer facing many sellers; e.g., Amazon) is not illegal.* I've read in several places that Section 2 of the Sherman Antitrust Act does not outlaw monopsonistic practices unless they are acquired by exclusionary conduct. There's not a lot of settled case law about what sorts of conduct are considered exclusionary by a goods retailer, as opposed to an employer. Future cases may change this, but it's going to be a near-vertical climb.

4. *Virtually all recent technology works in Amazon's favor.* Ebook readers, cheap tablets, ubiquitous broadband, POD machines, robotic sales data collection, online reviews, you name it: Amazon has almost no legacy baggage.

5. *Almost everything works against print publishing generally*, and the Big Five in particular. I'll come back to this.

I'm still not entirely sure how I feel about the whole business. In general, letting publishers set their own prices via agency agreements with retailers is a good thing because it allows startups to undercut them. The value to the public of any individual publisher (or conglomerate) is low, as long as startups have access to markets and can replace them. Access to bricks'n'mortar retail shelves has always been and still is tricky. Access to other retail channels has never been easier. If I were ten years younger I might be tempted to try again.

Now, why is Big Print in such trouble? Somebody could write a book (and I wish Mike Shatzkin in particular would) but here are some hints:

1. *Trade book print publishing is a big-stakes wager against public taste.* It's hard to predict what the public will want even in categories like tech. Literary fiction? Egad. Guess wrong, and you've lost what might be a million-dollar advance plus the full cost of the press run and any promotional efforts.

2. *The economics of trade book publishing are diabolical.* Trade books are basically sold on consignment, and can be returned by the retailer at any time for a full refund. This makes revenue projection a very gnarly business. Books assumed to be sold may not *stay* sold.

3. *Online used bookselling reduces hardcover sales.* Buying a hardcover bestseller soon after release is a sort of impatience tax. The impatient recover some of the tax by listing the book on eBay or Amazon Marketplace at half the cover price. The patient get a basically new book for half-off, and then sometimes sell it again…for half the cover price. This would not be possible if online searches of used book inventory weren't fast and easy.

4. Related to the above: *Remaindering teaches the public that new hardcovers are cheap.* Most print books are eventually remaindered. The remainders are generally sold online for as little as three or four bucks. They're new old stock books with a marker swipe on one edge. The more publishers guess wrong about press runs (see Point #1) the more books are remaindered, and the more hardcovers lose their mystique and (more important) their price point.

5. *Fixed costs for the Big Five are…Big.* There is a very strong sort of "Manhattan culture" in trade book publishing. Big publishers are generally in very big, very expensive cities, which carry high premiums for office space and personnel. My experience in book publishing suggests that none of that is necessary, but as with Silicon Valley, it's a cultural assumption that You Have To Be There, whatever it costs.

Bottom line: The Big Five need the $25 (and up) hardcover price point to maintain the business model they've been evolving for 75 years. If hardcover sales ramp down, they need ebook sales to make up the difference. Ebooks are cheaper to produce and manage (i.e., no print/bind costs, shipping, warehousing, or returns) and it's quite possible that a $20 ebook price point could stand in for a $30 hardcover price point. However, Amazon has trained the public to feel that an ebook shouldn't cost more than $10. Indies have put downward pressure on even that, and the demystification of hardcovers via used and remainder sales hasn't helped.

What options do the Big Five have? Culture is strong: They're not going to cut the glitz and get the hell out of Manhattan. (That may not be invariably true; Wiley US moved from Manhattan to New Jersey some years ago. Wiley, however, does not publish trade fiction and has never been deep into glitz. I doubt, furthermore, that they would ever have moved to Omaha.) A reliable midlist might help, but midlist titles now exist mostly as ebooks. Most publishers, big and small, have long since outsourced design and production to third parties, and are already doing a great deal of printing—especially color—in China. Beyond that, I just don't know.

Don't misunderstand: My sympathies are with publishers, if not specifically large publishers. I was in the trenches and I know how it works. Books can only be made so cheap before quality suffers, especially ambitious nonfiction like Steven Pinker's *The Better Angels of Our Nature*. We may be in a race-to-the-bottom that cannot be won by either side. What I'd really like is honesty in all quarters about the issues and (especially) the consequences. Rah-rah tribalism helps no one.

Both sides have points in their favor. Amazon has done something not well-appreciated: It's made it possible for self-publishers and indie publishers to reach readers. Physical bookstores have long been barriers to entry in publishing. Quality remains a problem, but hey, is that a *new* problem? Traditional publishers claim that they guarantee quality, even though "quality" is a very tough thing to define. Most of my life I've abandoned a fair number of print books every year as unreadable, not because I dislike the approach or the topic but because the writing is bad. This is supposedly the value that publishers add. The adding is, shall we say, uneven.

My suggestions sound a little bit banal, even to me:

• Publishers need to pay more attention to objective quality. Bad writing is a fixable problem; you either don't buy it, or you fix it after you buy it if you judge the work important enough to go forward. This is the edge traditional publishers have over the indies.

• Amazon needs to consider that book publishing is an ecosystem in which many players have important roles. Market share won't matter if you kill huge segments of the market. Amazon may simply not care; there's plenty of money in selling thumb drives and diapers.

• Readers need to meditate on the realities of writing. Writers need to be paid. Cover price isn't everything. Quality *matters*.

• The hardcover as the core of trade publishing must die. Hardcovers need to become a luxury option. If I read an ebook or paperback of a truly excellent work, I may want a hardcover, and we're very close to having the machinery to do hardcover onesies at reasonable cost. I've upgraded to hardcover many times, but generally on the used market, since by the time I read a paperback the hardcover may already have been remaindered and unavailable new.

• Publishers need to ask themselves if Manhattan and San Francisco really deliver benefits comcomitant to their astronomical cost.

• Amazon is a given. The Internet leans toward channel capture. If it weren't Amazon it would be someone else. Grumble though we might, we need to start there and figure out the best way forward.

In the meantime, remember: There are countless sides to every argument, and no easy answers to anything. *You are always wrong*. And so am I. Get used to it.

All the Forks That We Need

From Jeff Duntemann's Contrapositive Diary, October 17, 2009

Carol and I have been married now for 33 years. Back in the summer of 1976 my mother threw us a bridal shower, and among the many gifts we received were two sets of Ecko Eterna Corsair stainless steel flatware, for a total of eight place settings. We still have them. In fact, we have been eating with them for all 33 of those years. (Below left is a 33-year-old daily-driver fork. "Eterna" is fersure.) They're all still in the drawer.

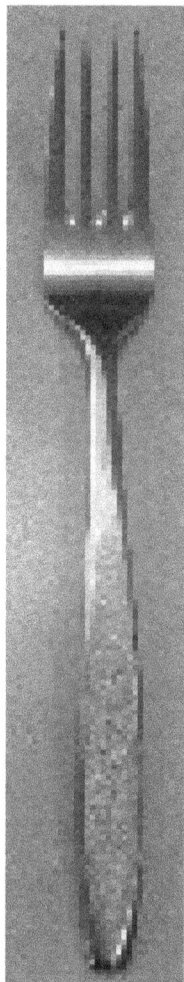

Well, almost all of them. Flatware eventually goes missing like protons, though with a much shorter half-life. Over the years a couple of spoons and forks have probably followed us to pot-lucks and never come home. I have no better explanation. When I was a toddler I used to drop flatware down the cold air return, which I know because when I was 14 I helped my father tear out the old sheet-metal octopus that heated our house, and found most of a place setting at the bottom of the big pipe. As an adult I have no such excuse. I only know that we run out of clean forks before we run out of clean tablespoons.

I got irritated enough recently by our fork shortage to look on eBay, where I scored three Ecko Corsair forks for $10–and five spoons for $12. The forks were unused, and when I got them, washed them, and dropped them in the drawer, it struck me that there wasn't much difference in appearance between the brand-new Corsair forks and the forks that have been faithfully stabbing our steaks for 33 years now. We have a full drawer of flatware again, and all the forks that we need. Better still, if we ever need more, we know where to find them.

I had an insight when the forks arrived that Carol and I are not and will probably never again be in the market for new-build stainless steel flatware. Why should we be? Our set works perfectly, and still looks like new. Spare parts are available, cheap. This isn't good news…if you make flatware.

And I also wonder if our auto industry is in trouble at least in part because cars are lasting longer and people are trading them in far less often. I got my first car in 1970 when I started college. It was a bare-bones 1968 Chevelle 300, and even at two years old the door panels were growing significant rust spots. By 1974 the body was mostly rot and the engine disintegrating, and rather than pony up for a valve and ring job, I dumped it and bought a brand-new Honda Civic. The Civic lasted until 1982, when its brake cylinders started going out repeatedly. I had a Datsun pickup for a year and decided I didn't like pickups; I traded it for a 1984 Chrysler minivan, which I owned uneventfully until 1995. That year I traded the old minivan in on the newest version of the same minivan–*and we still have it*, a little tired but entirely functional. [Author's note: We finally traded the van in on a Dodge Durango in 2014.] The Toyota 4Runner that we bought in 2001 will flip over 100,000 miles today or tomorrow, and has never given us a lick of trouble. No rust, no wiggles, no funny noises, no problemo nada. I expect to be driving it happily ten years from now. [Author's Note: We still have it in May 2021]

Draw the curve here. Cars that used to implode after 5 years are now lasting for fifteen or more. Is it any wonder that we don't need as many cars as we used to? A great many of our economic problems today may stem from simple over-capacity: factories cranking out stuff like it's 1968, simply because that's what they've always done and the spreadsheeters require it. (Publishing certainly has that problem, though for different reasons.) We are the victims of our own success, in that there is less work than there are workers, because we're making better forks…and *much* better cars. We may not need a Big Three for making cars. A Big Two may be sufficient. (I'll leave the eenie meenie mynie moe part to someone else, thanks.) And if that's the case, we have to be *extremely* careful about protectionist economics, because the export market is all that's left, once Americans have all the forks that they need.

What Dogs Gave Us

From *Jeff Duntemann's Contrapositive Diary*, July 30, 2010

We domesticated dogs. And dogs, in return, made human civilization possible.

Work with me here. A lot of my recent reading has been about human origins, stemming from my fascination with *Homo Neanderthalis* and what became of him. Two books of note: *The Third Chimpanzee* by Jared Diamond (1993) and *Before the Dawn* by Nicholas Wade (2007.) Jared Diamond is always a good read, and even though the book is showing its age I strongly recommend it. Wade covers much of the same turf, but does so with the tools of DNA analysis that simply didn't exist twenty years ago, when Diamond was doing his research. By counting mutations and working backwards through Y (male) chromosomal DNA and mitochondrial (female) chromosomal DNA, we can infer a great deal about human populations, where they came from, how they changed, and when. Of some of it I'm dubious–the extrapolation about the sources of human language, for example, seems a stretch–but most of it is no longer controversial, nor even exotic.

Both authors draw on anthropological research of stone-age peoples who survived into the 20th century. (Diamond did a lot of that research himself, in New Guinea.) The picture they paint of early humanity is grim: We are not fallen angels. We are risen apes. The hallmark of early humanity was deliberate genocide: New Guinea tribesmen told Diamond straight-out that their overall tribal goal was the extinction of other tribes. The homicide rates among such tribes are many times that of the homicide rate in Detroit; men who cannot claim to have killed another man often cannot persuade women to marry them. This seems to have been the pattern for hunter-gatherer societies as far back as we can see via the fossil record. Many Neanderthal skeletons show the marks of multiple healed bone and skull fractures, and a couple of them evidence of spear impingement on bone. Constant warfare was the pattern, and the method (judging from modern stone-age peoples) was the dawn raid: Raiders would stealthily draw close to a rival tribe's encampment, and wait for the rivals to turn in. Then, when there was just enough dawn light to move well, the attackers would fall upon the sleeping rivals and spear them where they lay.

This worked, and worked well. People have to sleep, so the attackers had the advantage. Then one day about 15,000 years ago, something unexpected happened: Animals around the rival encampment sensed the attackers creeping in for the kill, and set up a huge and unfamiliar racket. The rival group, awakened by the animals, grabbed their spears and gave chase. The attackers had been up all night waiting for just the right moment. The defenders had just had a good night's sleep. They could outrun their sleepy-eyed assailants, who had a ways to go to return to their home turf. More than a few attackers probably took a spear through an eye socket, and once enough of your dawn raiders take a spear through an eye socket, dawn raiding becomes a *lot* less compelling.

All because of some previously unknown animals who looked like wolves but made noises that wolves did not make–and appeared to consider the rival camp to be friends rather than food.

As best we can tell, dogs were first domesticated about 15,000 years ago, which was just about the time that *Homo Sapiens* was moving from wandering hunter-gatherer societies to settled societies that eventually became agricultural and pastoral societies. Just how they were domesticated is still unknown, but the work of Belaev and his silver fox suggests simple selection by temperament: Ancient wolves became camp followers, and ancient humans tossed them scraps. Wolves who could stand to be near humans ate better without working as hard and had more pups. The few stone-age tribes we've been able to study sometimes captured wild animal juveniles and kept them as entertainment until they became grouchy on maturity. Dogs need to be handled as puppies to be fully at peace with humanity as adults; perhaps those wolves-in-transition descended from adult wolves who were handled by humans as pups and remembered: *Those two-legged whatchamacallits handled me without hurting me–and they toss me aurochs bones!*

15,000 years ago, that was a helluva deal if you were a wolf.

Explaining the bark is tougher, but group selection suggests that if some quirk in the genes of certain wolves allowed those two-legged whatchamacallits to survive and thrive, there'd be more aurochs bones and more yappy wolf/dog pups. Evolution works fast: Belyaev turned wild fox into peculiar (if not completely domesticated) pets in only 40 years, simply by selecting fox who were most willing to be handled when young and least snarly and aggressive when mature. A fox who will lick your face instead of biting your nose off is most of the way to a dog anyway; in another hundred years, he'd be sleeping at the foot of your bed and fetching tennis balls.

The bottom line is this: Without dawn raids, settled living rather than wandering became possible, and settled living fostered the development of villages and agriculture and trade and writing and all the other precursors of the lives we live today.

The Neanderthals had bigger brains than we do. *What they didn't have were dogs*. And, lacking dogs, the unfortunate louts dawn-raided one another to extinction, leaving *homo sap* and his faithful yappers to pick up the turf and eventually take over the world.

Raise a glass of Laughing Lab Ale to *canis familiaris*: Everything we are we owe to him. *Good* dog!

The War That Nobody Dares Explain

From *Jeff Duntemann's Contrapositive Diary*, November 11, 2018

Armistice Day. I call it that in this entry because 100 years ago today, The Great War (now called World War I) ended. We've broadened the holiday to all those who have served in war on our behalf, but until 1954, the day was named after the armistice that ended WWI.

My grandfather Harry Duntemann served in The Great War. I never got to talk to him about it because he died when I was four, or I would have asked him what caused the War. I'm not entirely sure he could have told me. Degreed historians have been unable to tell me. I've read a pile of books about it, but as close as I've come to an answer is simply that Europe's leaders were about ready for a war, and when the assassination of a second-shelf political figure provided them with an excuse, they went for it. Four years and sixteen million deaths later, the armistice was signed, Europe was rearranged, Germany thoroughly humiliated, and all the pieces put in place for an even greater war a generation later.

Bad idea, top to bottom.

Here's my theory, which I offer as speculation based on a view from a height: WWI was a pissy argument among Europe's ruling elite, made deadly by industrialization and technologies that hadn't been dreamed of during the Franco-Prussian War in 1870. Certain members of this insufferable boys' club took offense at other members' reaction to crackpot Princip's terrorist attack, leading to others taking offense at their offense, leading to a wholesale loss of face among the elites, who threw the inevitable tantrum and leveled half of Europe in the process. They're still digging up live ordnance in places a century later. Lots of it. Sometimes it explodes. In terms of casualties, WWI has never really ended.

The common element here? Inbred ruling classes who cannot conceive of being wrong about anything. In 1914, they were elite by virtue of aristocratic birth, or sometimes having risen through the local equivalent of civil service. That era was the transition from "the King can do no wrong" to "the government can do no wrong," which was perhaps a step in the right direction, but…

...we still have ruling classes, and they are *dangerous*. Graduate from an Ivy and you're set for life. Along with the diploma you're given the impression that you're just...*better*...than people who go to state schools, or who eschew college altogether. This leads to a pathological inability to doubt your own view of the universe, and in most cases, your own expertise. Given too much power, such people can, have, and will continue to destroy entire nations.

Self-doubt is an essential personality trait. I consider it the single most reliable indicator of people who are high in both rational and emotional intelligence. A modest amount of self-doubt among Europe's elites could have stopped WWI. Stopping WWI, furthermore, might well have stopped WWII.

I don't honestly know what one can do about ruling classes. Not supporting political parties would be a good first step, because political parties are mechanisms that make the elites rich and keep them in power. But you know how likely *that* is. Redistribuing power (not wealth) would be another good step. Again, this would mean broadening access to the Ivies (ideally by some sort of entrance lottery) and limiting the powers of government far beyond the degree to which government would allow itself to be limited. (I have a good political novel on the subject in my notes that I won't write, because political novels are depressing.)

And even that might not work. Once again, we run up against the primal emotion of tribalism, from which most of our current troubles emerge. That's a separate topic, but not an unrelated one.

My advice? 1. Shun the ruling classes. You'll never be one of them (no matter how much you think you deserve to be) and fostering ordinary people's desire to be among the elites is how the elites keep ordinary people under their control. 2. Limit government power at every opportunity. The less power our elites have, the less damage they can do. 3. Read history. Granted, I read a lot of history about WWI and still doubt whatever understanding I thought I gleaned from those books...

...but let me tell you, I understand the Jacobin mindset *completely*.

Lots of Supermarkets

From *Jeff Duntemann's Contrapositive Diary*, February 3, 2016

Twenty-odd years ago I remember reading a compendium of "real-world" ghost anecdotes. They weren't stories, just individual reports from ordinary people who were not looking for ghosts but ran into them anyway. One of my favorites was a report from a widow in England who saw her recently deceased husband on the staircase every night for a week. The man looked happy, but said nothing until his final appearance, when he spoke one sentence: "There are lots of supermarkets where I live." Then he winked out and she never saw him again.

Well. I can think of a lot of better things to tell your grieving spouse when you appear to them postmortem:

- I'm all right.

- I love you.

- I forgive you.

- God is good.

- There is $10,000 in hundreds stuffed inside the living room couch.

But…lots of supermarkets in heaven? That is so unutterably weird that it lends credence to the report. Why would the widow make something like that up?

Maybe she didn't. My experience here in Phoenix for the last month and a half suggests that it may not be so weird after all. Work with me here: Until six weeks ago, Carol and I lived on the slopes of Cheyenne Mountain near a town of about 400,000 people. Colorado Springs is not a small town, but we still had to drive 75 miles to Denver for certain things, like The Container Store and any useful bookstore that wasn't Barnes & Noble. Today we live in America's 6th largest city (instead of its 41st largest city) and if you toss in suburbs like Mesa and Scottsdale, the metro area has four and a half million residents.

Nor are we way out on the fringes of things, like we were when we lived in Cave Creek in the 1990s. We're right down in the thick of it all, three blocks from tony Scottsdale and a little over a mile from the Kierland neighborhood, where the primary occupation is spending money by the livingroom couchful.

The amount of retail here is staggering, as is the number and sheer diversity of restaurants. I didn't know that Mexican Asian food was a thing, but it is, albeit what sort of thing I'm not yet sure. (When I decide to find out, well, it's just a few miles down Scottsdale Road.) Driving around the area, Carol and I go into a sort of Stendhal syndrome trance at times, boggling at the nose-to-tail storefronts and shopping centers within a couple of miles of us. It's not like we're hicks from the sticks; Colorado Springs is hardly the sticks. But we've never seen anything even remotely like it.

There is a supermarket called Fry's Marketplace a few miles from us that is about twice the size of any other supermarket I've ever been in. They have a wine bar, a sushi bar, a substantial wine section (something we didn't get in Colorado due to corrupt politics) and plenty of stuff that may or may not be appropriate for selling in grocery stores, like…livingroom couches. (Eminently stuffable ones, too.) Outside there's covered parking and a car wash. Oh, and valet parking if you don't want to walk in from the far corners of the lot.

Now…what if we *were* hicks from the sticks?

I wager that we'd pass out in astonishment. Yes, I know, we all get lectured a lot about how we shouldn't obsess on material goods. So who's obsessing? I think I come out better on this score than a lot of people; granted that I hoard variable capacitors and never met a radio tube I didn't like, absent the occasional gassy 6AL5. Read this twice: *There is a huge difference between wanting everything you see and seeing everything you want.* I don't want all that much, but I appreciate being able to get things that I do want, weird or uncommon though they might be.

I can empathize with that poor old dead guy in England somewhere. Perhaps he lived all his life in a village in Cornwall, and ate the same things all the time because the same things were all there were in his village. Maybe he was poor. Maybe he just got damned sick and tired of bubble and squeak. He knew the world was a richer place somewhere, but his own circumstances didn't allow him to get there.

Then his heart gives out, and wham! God drops him out in front of some heavenly Fry's Marketplace, where your credit cards have no limit and you never have to pay them off. (Maybe he met Boris Yeltsin there.) Good food, lots of it, and never the same thing twice? That could be all the heaven some people might want. I think I understand why he came back to tell his wife about it.

So. Like most people, my collection of loathings has swelled as I've passed through middle age. I don't like green vegetables, and haven't now for 63 years and change. Along the way I've picked up loathings for certain philosophies and people, like Marxism, Communism, and the sort of virtue-signaling wealthy socialistic urban elitist busybodies who buy $59 titanium pancake flippers and then wear torn jeans to show their solidarity with the working poor.

Far worse are the people who assume that their way is the *right* way, and that if I don't see things their way, well, I'm a [something]-ist and deserve to be re-educated in the gulag of their choice.

Choice, heh. Choice is a good word. Freedom means choice. Choice does not mean overconsuming. Choice means being free to consume what I want, and not what some worthless meddling government apparatchik thinks I should want. I walked into Fry's Marketplace. It was a wonderland. I walked out with a smile on my face and a bag of gemstone potatoes under my arm. That, my friends, is America. Slander it at your peril, and ideally somewhere out of earshot of the rest of us.

Review: William Banting's
A Letter on Corpulence

From *Jeff Duntemann's Contrapositive Diary*, September 16, 2009

Do you like Banting?" "I don't know. I've never banted."

Unlike the oft-quoted line about our man Rudyard, this isn't really a joke. I have banted, I'm still banting, and I do like it. However, I didn't know it had a name until a couple of months ago, when I read William Banting's *A Letter on Corpulence, Addressed to the Public,* and began to research the booklet's background.

Dr. Atkins, shove over. Mr. Banting was here first.

In London in the early 1860s, an overweight undertaker was talking to doctors about his obesity. He had watched himself put on weight over the previous thirty years until, at age 65, he weighed 202 pounds, and stood only five foot five inches. He was having trouble getting up and down stairs and doing simple things like tying his shoes. He was annoyed. He had tried everything local physicians suggested, including buying a boat to row on the Thames and walking briskly every day, and taking various medicines that we would today consider worthless nostrums. Nothing worked. Then he came upon Dr. William Harvey, who made a suggestion that seemed too simple to be useful: give up beer, sugar, and "farinaceous" (starchy) foods.

Banting did so, beginning in September, 1862. And fortunately for us, he was of a scientific turn of mind, and wrote down both what he ate daily, and what he weighed every three weeks, for the following year. And in that year he dropped 46 pounds, eating mostly meat and non-starchy vegetables, plus a piece of dry toast or rusk (zweiback) for tea. And he lost the weight even eating four meals a day and drinking an amount of alcohol that would leave me unconscious on the floor.

After losing about a pound a week for that year, he felt better than he had in two decades, could navigate stairs without hyperventilating, and do whatever he needed to do in terms of ordinary activities. He felt that his eyesight and hearing had improved. He was, in short, a happy guy. And having achieved his goal of losing significant weight, he did a remarkable thing: He wrote up his experience as a pamphlet addressed to the public (what today we'd call an "open letter"), printed it at his own expense, and then handed it out to anyone who was interested.

It was popular enough to warrant two sizeable addenda across several printings, but even with those included the whole thing is only 25 pages long, and available as a free facsimile scan from Google Books. You can read it in fifteen minutes, though people who are not used to Victorian diction may find the text a bit of a slog. The pamphlet became popular and was much discussed in the London area at that time, enough so that "to bant" became a new verb, and meant to adopt Banting's diet as a means of losing weight.

The Google Books edition include two longish contemporary commentaries, one from *Blackwood's Magazine*, the other from *Harper's Weekly*. Both are snarky wanders intent on demeaning Banting's experience, and neither confronts the truth face-on: Banting did an experiment, recorded his results, and made them public without any attempt to profit from them. (In fact, he gave 50 pounds to a local charity hospital in thanks.) Instead, *Blackwood's* tries to convince its readers that Banting was not all that fat to begin with, and besides, fat people tend to be affable and law-abiding citizens, so it's *good* to be fat! There's not a lot to be taken away from the two reviews except the sense that things don't change much; many of the same groundless arguments are thrown today at low-carb diets, simply because "everybody knows" that eating fat makes you fat and the best course is a "balanced diet," which, as always, means "a diet that I favor."

William Banting is important because his experience predates the modern carb wars by close to a century. He wasn't trying to debunk Ancel Keys' fraudulent research or establish a diet-book empire. He was just writing down something that had worked for him, and he cautiously suggested that, under advice from their own physicians, overweight people might try the same method. It may not work for everyone, but (in contradiction to the ridiculous critique in *Blackwood's*) that does not mean it will not work for anyone.

Highly recommended, especially since you can read it over your eggs and bacon at tomorrow's breakfast.

The Limitations of Celebration

From *Jeff Duntemann's Contrapositive Diary*, January 5, 2010

Yesterday saw an annual ritual performed here: Popping the Christmas mix CDs out of the 4Runner's CD changer, and dropping back in four non-Christmas selections from the 20-odd mix CDs I have in the CD wallet. Tomorrow is the Feast of the Epiphany, which, in the minds of most Catholics–my own included–is the end of the Christmas season. Our tree is drying out, and probably Thursday or Friday we'll pull down and pack the ornaments, re-roll the 10 strings of old-style lights, and get everything back into its proper box. I'll run the Lionel trains around the track one more time and pronounce them good, as I have now for 30-odd years of Christmas railroading. The tree will come down and be taken to Rocky Top, a local rock-and-dirt yard that recycles Christmas trees into mulch. The trains and all the decorations will go back down under the big stairs in the Harry Potter closet, to wait quietly for Christmas 2010.

It wasn't that I was getting tired of the Christmas mix CDs, or the tree, or the decorations, or (especially) the Lionel trains. Far from it. What I don't want is for them to become *ordinary*. That is in fact the risk of endless Christmas celebration, as it sometimes seems to lean in this country. Nor is it specifically a Christmas problem: Celebrate *anything* too much, and whatever you're celebrating merges into the landscape of ordinary life, and loses its power to remind, to re-orient, to refresh.

It's important to separate the mechanics of celebration with what is being celebrated. The *ideas* behind Christmas, whether the purely religious or the secular (there's nothing essentially religious about "peace on Earth, good will toward men") are worth keeping close at hand at all times of the year, and ideally through every waking moment we live. The tree, the lights, the egg nog, the creche, and all the rest of it serve as reminders, popping up on an annual basis to reinforce the importance of keeping generosity, civility, and patience in our everyday toolkits. If we engage in the mechanics of celebration for too long, the symbols lose their power, and because symbols demand attention, they can too easily become irritations.

So it's essential to the mission of Christmas to put it all away after awhile, lest we get all Christmased out and cease to see the point of it. Come next fall (and a little later than November 1, perhaps? Please?) we'll begin to remember again, and the symbols will re-emerge, refreshed in their power just as we left the previous Christmas season refreshed in our convictions. We'll rediscover the delights of Christmas music and Lionel trains and colored lights all over the place. We'll remember why we do it, and resolve to keep the ideals in our minds as we slog through yet another damned miserable winter.

It's time. I'll miss the trains (and maybe the egg nog) but yes indeed, it's time.

10,000 Pirated Ebooks

From *Jeff Duntemann's Contrapositive Diary*, December 29, 2009

Ebook-related items have been gathering in my notefile lately, and this is a good time to begin spilling them out where we can all see them. The triggering incident was a note from the Jolly Pirate [an anonymous correspondent of mine back circa 2008-2010] telling me that one of my SF stories was present in a zipfile pirate ebook anthology that he had downloaded via BitTorrent. That people are passing around pirated versions of my stories is old news. "Drumlin Boiler" was posted on the P2P networks a few months after it was published in Asimov's in 2002, and my better-known shorts have popped up regularly since then. No, what induced a double-take was the name of the pirate anthology: "10,000 SciFi and Fantasy Ebooks."

10,000? You gotta be kidding!

But I'm not. Jolly sent me the 550K TOC text file, which is 9,700 lines long, with one title per line. Not all are book length, and many, in fact, are short stories. Still, the majority of all book-length SF titles I've read in the last thirty years are in there, and so was "Borovsky's Hollow Woman," albeit not under my byline. (I wrote the story with Nancy Kress, who is listed as sole author.) The only significant authors I looked for but did not find were George O. Smith and Charles Platt. (One howler: *Bored of the Rings* is said to be by J. R. R. Tolkien. Urrrrp.)

The collection is 4 GB in size. The Jolly Pirate said that he had downloaded it in just under three hours. He attached the file for "Borovsky's Hollow Woman," which was a plain but accurate 57K PDF. Intriguingly, the date given under the title is January 28, 2002. The damned thing has evidently been kicking around for at least seven years, if perhaps not in its full 4 GB glory. This suggests that the anthology is not entirely ebook piracy but mostly print book piracy. ("Borovsky" was never released in ebook form.)

Some short comments:

• I verified the existence of the anthology from the Pirate Bay search engine. It really does exist. (So, evidently, does the Pirate Bay, which surprised me a little, considering recent efforts to take them down.)

- 10,000 ebooks do not take a great deal of space by today's standards. (Admittedly, better files with cover scans would be larger.) No one will think twice about a 4 GB download for size reasons, when 750 GB drives are going for $69.95.

- The PDF is ugly. The lines are far too wide for easy readability and (since this is not a tagged PDF) not reflowable. That said, I did not find a single OCR error.

- The Windows pathname of the text file from which the PDF was generated is shown at the top of every page. The pathname includes the full name of some clueless Dutch guy, from whose Mijn Documenten folder the file came. Ebook piracy clearly belongs to the common people, not some elite cabal skilfully dodging the **AA.

- I've used a scanner to rip a couple of print books (plus ten years of Carl & Jerry print stories) and it is horrible hard work. However, the anthology demonstrates that if print is a form of inadvertent DRM (which I have long thought) it is not a particularly strong one. After all, as Bruce Schneier has said about DRM systems generally, they only have to be broken *once*.

This last item is key. A printed book is a worst-case challenge for an ebook pirate. Compared to cutting off the binding and making sure the paper pages all feed straight through the scanner ADF and then fixing the inevitable OCR errors, stripping out an ebook's DRM is trivial. If ebook piracy is not yet a big deal, it isn't because it's difficult. It's still because reading ebooks is borderline painful. I may not be typical, but if I can buy a used copy of a recent hardcover of interest for $10 or less, I'll choose the hardcover rather than an ebook at any price. Sooner or later the readers will catch up to paper, and by then, well, we may see a 4 TB file called "10,000,000 Ebooks About Everything" on the file-sharing networks, and it won't even take an objectionable chunk of our 80 TB hard drives.

You think I'm kidding? Let's compare notes in five or six years.

How about 12? I took a look at the Pirate Bay today in 2021, and the big file is no longer there. In fact, relatively few ebooks are listed at all. I think ebooks are cheap enough so that the pirates don't bother with them much. The big deal is movies and TV shows. Large-scale piracy has gone underground, to torrentboxes and private trackers. Victory? Sure.

An EBook Fate Worse Than Piracy

From Jeff Duntemann's Contrapositive Diary, February 7, 2010

I've had it in mind for some months now to conduct and publish an interview with a backchannel correspondent of mine who calls himself The Jolly Pirate. That's unlikely; Jolly didn't like the idea much, and more to the point, he doesn't pay much attention to pirated ebooks. He is not an ubercracker from the Scene and doesn't want to be. He knows where the stuff is and he downloads it. He doesn't upload at all, except for the uploading inherent in torrent downloading.

His motivation is interesting; from a height I'd describe him as a hoarder who downloads all sorts of things under the assumption that they may eventually be harder to come by. He's read a few computer books downloaded from Usenet, all of them .chm files, and treats them like a sort of third-party help system for the technologies he's interested in. The thing that makes me grin a little is this: He says he has over 50,000 ebook files on his hard drive, but he doesn't own an ebook reader. He doesn't read for fun and I get the impression that he doesn't read much at all unless he has to. I asked him why he downloaded all those books, and his answer was simple and obvious: "Because it was easy." Most of you have seen my entry for December 29, 2009. Jolly downloaded 10,000 ebooks in a couple of hours. That scares some authors and publishers a lot, and I'm still trying to get my head around the question. Tim O'Reilly said somewhere that piracy is like a progressive tax on success, and that's a useful metaphor. I rarely see my own material in the pirate channels. That is not true of Steven King or Ms. Rowling.

And in truth, something else makes me worry more than piracy. This isn't an original insight, though I don't recall where I first read it, but a major threat to success in writing today is the competition from *books that have already been published.* There are only so many hours in a life, and with most any popular print book available used but in good shape on ABEBooks for $5 or less, a given consumer never has to buy a new book at all, especially fiction. It's less true in nonfiction covering emerging issues and technologies, but for last year's news and mature technologies it's operative: All the Windows XP books that the world needs have already been published, and you can get most of them from the penny sellers for the (slightly padded) cost of shipping.

My point: *Existing books compete for reader chair-time with new books.* An enormous number of books have been published in the past twenty years or so, and that's not old enough for them to crumble into shreds. (Alas, my '60s MM paperbacks are doing exactly that, reminding me constantly what "pulp" means.) They're all still kicking around the used and unused remainder market, and will be for decades to come. All the arguing about ebook pricing that I've seen so far seems to ignore the fact that new books of either type compete with used print books, and ubiquitous Web access makes finding *precisely* what you want almost effortless.

Paying $15 for an ebook edition of a popular hardcover is a sort of impatience tax. Wait a few months, and used copies of the hardcover will be on ABEBooks for $5 or (probably) less, *including shipping.* Good books, too. If Big Media ever truly embraces ebooks, it will be as a means of defeating the Doctrine of First Sale and eliminating the used book market. (The legal issues there are still very much in play. Expect much agitation in coming years for new laws forbidding the resale of "used" electronic files.)

This shines some different light on the difficulties Google has had getting authors to sign on to the Google Books settlement. I'm not sure that all authors and (especially) publishers even *want* the orphan copyright issue to be settled. If it is, suddenly the Google scanning machine will drop what may eventually be hundreds of thousands of additional ebooks into the marketplace, all of them competing for quality chair time with whatever current authors are writing. That may explain why I've had so much trouble getting SF publishers to talk to me. People may not be reading less these days, but they're certainly reading and re-reading things that already exist. The value of what I write now is correspondingly less.

When a pulp becomes an ebook, it becomes eternal. Don't tell me about death due to storage or container format obsolescence. I still have SF copy I wrote using CP/M WordStar in 1979 and stored to 8″ floppies, now safely on a USB thumb drive in .rtf format. If USB ever becomes obsolete, all my files will follow me to whatever comes next–and will probably take five seconds or less to transfer.

There will always be a reliable if modest supply of book crazies and loyal fans who will pay top dollar for The Latest. Beyond that, market cruelties come into play that will make it a lot harder to break into the writing business for the forseeable future.

Piracy? What's that again?

To Be a Free Man

From *Jeff Duntemann's Contrapositive Diary*, June 29, 2012

Sixty today. In answer to the quintessential American question, *What do you want for your birthday?* I have a fairly simple answer: *to be a free man.*

That bears some explanation. I'm not looking for freedom from responsibility. I'm a man who keeps promises. Neither am I looking for someplace to swing my arms around irrespective of local noses. That's an ego thing, made difficult by the fact that most people don't always know where their noses are. No, what I want is a higher and much tougher thing: to be sure that the view of the world that I hold is *mine*, and not handed down from someone else.

That kind of freedom, like keeping in shape, requires regular exercise and attention to detail. Five years ago I described some of the great lessons of middle age, which mostly cook down to *Life is messy—and messy is good.* Since then, a great deal of my research has been devoted to the problem of freedom of thought, and freedom's great enemy, tribalism. I've watched in dismay as people I've known sometimes for decades have sold their souls for a nickel to various tribes, becoming tribal footsoldiers who reflexively hurl hatred and abuse at anyone and anything that dares disagree with their tribal owners.

The thing that really makes me shudder is that it appears to be primal behavior, bubbling up from our deeper and more apelike minds with little conscious intent. Much research suggests that we evolved within a sort of Big Man social model in which a few alpha men at the top (yes, almost always men) and their cronies enjoy power, resources, and sexual opportunity handed up from those beneath them in the hierarchy. At the bottom are the expendable footsoldiers who fling pointless insults at the other tribes while never realizing that the poo-flinging is a distraction designed to hide the fact that the game is being played at their expense.

This might be the subject of a couple of very interesting entries, if I could work up the courage to write them. If I did, I'm sure I'd get plenty of abuse myself. Pressure to conform to received opinion is intense, and the social networking revolution appears to have been designed specifically to facilitate the enforcement of tribal dogma. Most of the bumper-sticker entries I see on Facebook are precisely this: Like and share—or you're not really a loyal party member, eh?

No thanks. I'm a contrarian, which means that I question all wisdom that somebody attempts to shove in my face, and the harder they shove, the harder I question. I do have opinions, which I refine constantly and change when it's clear that they need changing. I talk about some of them here, when I can make an interesting point. But I am training myself to offer the courtesy of not attempting to convert others to my views, in the hope that the courtesy will be returned. For example, I have a religion. If you want to know more about Old Catholicism (and the insanely optimistic little corner of it that I inhabit) I'll be happy to tell you. I will not attempt to convert you, and I promise to respect whatever religion (or none) that you hold. As for the rest, I'm digging around just like everyone else. Sometimes I turn up a useful insight. When I do, you're likely to see it here. Most of the time I'm just digging, because digging is where insights come from. Alas, there's always more dirt than insight, and this is not a new problem, nor one unique to me. *Don't fling the dirt when you should be offering others the insights.*

As for turning 60, well, heh. It's a good round number, and easy to remember. As I've said over and over again, I already have damned near everything a man might want: a spouse who is intelligent, interesting, loyal, and drop-dead gorgeous to boot; a house and a chair and a good bright reading lamp; friends who challenge and surprise; dogs and tools and tube sockets in abundance, and a front-row seat at the greatest show human history has ever put on offer. The challenge now is to keep what I already have, including the discipline of being my own man without minimizing the right of others to be theirs.

I have it. I intend to keep it. This is a happy birthday indeed.

Taps

From *Jeff Duntemann's Contrapositive Diary*, May 27, 2013

Our house is on the slopes of Cheyenne Mountain, and our back deck looks down on Fort Carson. We hear bugle calls at 6:30 AM, noon, and 5:00 PM, accompanied by a cannon. Then every night at 10 PM, taps drifts up the mountain. If I'm still up and around, I go out on the back deck and remember three men. When the echoes die away among the mountain canyons, I salute, and whisper, "Thank you, gentlemen."

All three served their country in World War II. Only two came back. One gave me life. One gave me his daughter's hand in marriage. And one…he died for you, me, and everybody else who thinks that freedom matters critically in the history of our world.

Frank W. Duntemann was attached to the AACS (Army Airways Communications System) and served in Italy and later, Africa. He claims that he slept through the bombardment of Monte Cassino, and having seen him sleep through my puppy Smoker's biting his ear and drawing blood, I can almost believe it. He sent Morse code on a Vibroplex bug, and could copy it in his head and pound it out on a typewriter as fast as anybody could send it. I could fill a book with the stories he told.

Steve Ostruszka was in the Navy, and served on the destroyer *USS Woolsey* (DD-437) until the War ended. Or so we think. Unlike my dad, Carol's dad simply refused to talk about the War. Nearly all of what we know of his days as an engine mechanic on the *Woolsey* came from his brother Ed, who also served on the ship.

Then there was Robert Williams, Jr. of Necedah, Wisconsin. He graduated from Necedah High School in 1942, and enlisted in the Navy immediately. At some point in 1942, he kissed his best girl good-bye, and got on a train for California to fly to Hawaii and then ship out on the Pacific. In 1944, he was a radio operator on a torpedo bomber, and his bomber went down in the ocean.

His best girl was my mother.

Like millions of other men and women in WWII and since, Bobby Williams gave his life to help ensure that others could keep theirs, and it was true of me in a very weird way. Absent his sacrifice, I wouldn't be writing this. I wouldn't "be" at all, which is a strange thing to meditate on. Now, the chain of contingencies leading to any individual life is long, and doesn't end until the moment the genetic die is cast. Nonetheless, I wish that he *had* come back. He loved my mother, and no one should die so horrible a death at 19 or 20.

I'll go out and listen for taps tonight, and I will thank them again. I will thank countless others for what they gave and for the lives that were changed or lost altogether. We must go to war with the utmost reluctance, but having done so, we must honor those we send, in the hope that every war will be the last that our world will ever see.

Frank W. Duntemann

Steve Ostruska

Robert Williams, Jr.

Dancing with Diction

From *Jeff Duntemann's Contrapositive Diary*, March 3, 2012

Today is the birthday of Dr. Seuss, without whom I would care nothing for poetry. One of the great bonding behaviors I shared with my baby sister was running around the house reciting snatches (sneeches?) of kid-book poetry at the tops of our lungs. "This one has a little star! This one has a little car! Say, what a lot of fish there are!" The king of that castle is and will always be Theodor Geisel 1904-1991. Circa 1960 our parents had signed us up for what amounted to the Dr. Seuss book club, and every month we got one of his books or another book that was clearly written in his style. There were some outliers not written in verse, like *Look Out for Pirates!* but who remembers those anymore? (*Go, Dog, Go!* may be one exception.)

On the other hand, I only have to recall the title of *One Fish, Two Fish, Red Fish, Blue Fish* and my poetry-reciter is off at a trot. Gretchen's even better at it than I. Don't get us started if you're one of those lit'ry types who feels that any poetry with rhyme and meter is worthy only of folding into the center of a Hallmark card.

Modern universities crank such out by the pallet load. Years 'n years ago, at one damned cocktail party conversation or another (I think associated with the Book Expo America trade show) I made an energetic case that good poetry can have both rhyme and meter. A well-credentialed tribalist immediately jumped on me, steam jetting from every orifice. "So," he jetted, "all poetry should be doggerel?"

Whoo-boy! Note the well-worn tribal tactic: I suggested that something the tribalist hated should be allowed. The tribalist immediately misrepresented me as saying that everything except what he hated should be forbidden. I called the fool on it. I basically humiliated him in front of several of his peers. How did I humiliate him? I dared him to begin reciting blank verse from some author who would be taught in college literature courses. He couldn't do it. I turned the knife by immediately beginning to recite "The Hollow Men." I stopped after eight or ten lines. I then asked him which poet had written the following:

mighty guest of merely me
—traveler from eternity;
in a single wish, receive
all I am and dream and have.

He shook his head. "You did." Heh. Don't I wish. It was e. e. cummings. I offered to recite the rest of the poem. The dork said "No thanks," and slunk away.

Now, I may be a better memorizer than he was. But I had a secret advantage: Structured poetry is easier to remember. And a secret vulnerability: I had recited all of Eliot that I could recall, and I remember Eliot today largely because I used to make fun of him so much. Give me Macavity any day, even if the sophisticates dismiss it as children's poetry. (It's a cat poem. Dare 'ya to call it *dog*gerel!) I can recite a great deal of that. It contains irony, subtlety, and much merriment. You can dance to it. I give it a 10.

Note that I don't "hate" blank verse and freeform poetry, nor do I dismiss it simply because it lacks rhyme and meter. I studied it. I studied Walt Whitman, Robert Lowell, Wallace Stevens, Theodore Roethke, and all those guys of that era and that school of poetry, which has basically won the day. I still recall why my profs thought they were significant. The problem is that the poems themselves I have utterly forgotten. Lowell has a great line somewhere about '59 Chevies rolling past like fish in a tank, "in finned servility." But that's all of him that I can remember, having read an entire book full of his stuff and discussed it at length in a 300-level class. I'm sure it was carefully crafted. I'll grant that it was important. But in no way on this or any other world could it ever be *fun*.

For that you have to go back to poets like Vachel Lindsay, who opened "The Santa Fe Trail" in an eminently memorable way:

This is the order of the music of the morning:
First from the far east comes but a crooning.
The crooning turns to a sunrise singing:
Hark to the calm-horn, balm-horn, psalm-horn.
Hark to the faint-horn, quaint-horn, saint-horn…

Damn, not only can I see that, I can *feel* it! It makes me want to run around the house with my baby sister (now 55) yelling *"Ho for Kansas land that restores us! When houses choke us and great books bore us!"* Eventually we collapse on the couch, breathless from laughing so hard and glowing from feeling so good. Kid stuff? Sure! At least for kids who haven't yet sold their kidness for a pot of message.

Poetry is about laughter, especially laughter that comes of wishing we could be in Kansas so that we could get away from all those Great Books that are so ponderously self-important they they undergo lexical collapse and vanish into their own navels while everybody stands around scratching their heads trying to understand what the hell they were attempting to convey.

And about dancing, yes. Poetry is dancing with diction, doing the polka with participles, spinning an allemand with adverbs. It's cutting loose from grim reality for awhile and letting language just *take* us. "He thought he saw an elephant / That practiced on a fife: / He looked again and found it was / A letter from his wife." What does it mean? Who knows? But who cares? It's *fun*.

If I'd had to jump straight into Lowell and Roethke I would have tossed it all overboard. But Dr. Seuss had gotten to me first. He taught me that you could dance to words, and from that dance it was a short step to Chaucer and Pope and Longfellow and Tennyson and Lindsay and Robert Frost and e. e. cummings. Having danced to the edges of rhyme and meter (cummings is a great transition) I could go the rest of the way, and watch the tailfins go by with Robert Lowell.

Did poetry classes leave a sour taste in your mouth? Grab a Dr. Seuss book, and find your sister if you have one. Run around the house spitting iambs and trochees until you collapse laughing on the couch. That's how you reboot your poetry sense. Then, if you want, you can take it all the rest of the way to Walt Whitman and beyond.

But I personally wouldn't blame you if you stopped right there.

Delores Ostruska 1924-2013

From *Jeff Duntemann's Contrapositive Diary,* February 4, 2013

Carol's mom has left us. She died quietly this past Saturday after a long illness, at a nursing facility near her Crystal Lake, Illinois home. Her daughter Kathy was by her bedside, and her two grandsons Brian and Matthew had visited her earlier that day. She was 88.

Most people in our time are lucky to have two loving parents. Somehow, incredibly, I had four. I met Delores on August 2, 1969, when I came by their house to pick up Carol for our first date. I was 17, a little scruffy, and undoubtedly, well, *odd.* No matter. Delores smiled and welcomed me, a welcome that never faded. Carol's dad was a slightly harder sell, but I won his esteem by treating his daughter with respect and kindness. When I bought a lathe in 1977 he stabled it in his basement, and over time he taught me what he knew about its use, which (considering that he could grind a carbide die to a ten thousandth of an inch accuracy) was pretty much everything.

On many Sundays Delores prepared family dinners for which her sisters Marie and Bernice and her Aunt Marie and Uncle John drove up from the South Side. Pork roast, salad, vegetables, bread, dessert; a huge spread brought to the table hot and perfect in all ways. I had a place at that table, as later on Kathy's boyfriend/fiance/husband Bob did as well. It was decades before I knew the term for the feeling that hovered all about us in Delores' dining room, but when I found it, many things fell into place. It was unconditional love.

I had had that from my own parents, of course. And even my own father was a bit of a hard sell, since I bore little resemblance to the rowdy boy that he himself had been and expected his own son to be. All the more remarkable that Delores and Steve embraced me almost immediately as one of their own.

Delores was a child of Polish-American heritage, youngest daughter of a large family, who was born and grew up on the Near South Side of Chicago. She belonged to a group of very close teen girlfriends who called themselves The Comets. They were capable and confident girls, journeying around the city for fun, and even slept on the sand of Chicago's 31st Street Beach. She quietly rejected the dour Polish pessimism of her own parish church, and far preferred the exuberant Catholic culture of an Irish parish a few blocks away. She believed all her life in an infinitely loving God and the goodness of all His creation. When I began struggling with my own life of faith at the dawn of middle age, it was her example that helped bring me to the unbounded and unshakable Catholic optimism that I hold today.

Delores worked at the US Treasury in downtown Chicago, where she helped trace lost and stolen US Savings Bonds. During WWII she met and in 1947 married Steve Ostruska, one of her brother Charlie's Navy shipmates. After Carol was born the family moved to Niles, Illinois, where Delores lived for over forty years before moving in with her daughter Kathy in Crystal Lake.

Every summer while the girls were small the family vacationed along the lakes near Hayward, Wisconsin, where Steve fished for walleye and city girl Delores learned to love the outdoors.

It's not fair to picture her as an elderly woman when she has already broken the bonds of this Earth and risen triumphantly into the arms of the God she so strongly believed in. I prefer to recall her as the beautiful, vigorous person she was most of her life. In truth, all the time I knew her she glowed with the quiet, invincible light of unconditional love, and if there's anything closer than that to the ineffable light of God, I don't expect to see it in this world.

Hypothesis: Reason vs. Anger

From *Jeff Duntemann's Contrapositive Diary*, January 28, 2013

I've been developing a hypothesis in the back of my head for some time now: That evolution developed anger as a countermeasure to reason.

That anger and reason are forces in opposition is obvious to anyone with an IQ over 75. That's not what I'm talking about. What I'm trying to explain is how anger came to be. And with so much else in my recent research, it all comes back to tribalism.

We evolved from killer apes, and from killer apes we inherited a peculiar but very effective survival mechanism: the tribe. Tribes are an interesting piece of biological machinery. They're actually genetic amplifiers for what we now call "alpha males," and the idea is to select for the genes of the meanest badasses in the neighborhood, so to better compete with the badasses living on the other side of that hill over there.

While we were killer apes and primitive hominids, it worked very well. Evolution is always trying new things, however, and a few tens of thousands of years ago something new appeared: abstract thinking. On an individual level it was a big win. Hominids who could think their way through a sticky situation would leave more children than hominids who just followed their killer ape instincts. But on a group level, it tended to erode the much older tribal mechanism. Let me demonstrate what I mean with thirty seconds of drama:

[Foot Soldier:] Boss, out on the front lines, we've been thinking. Ten of our guys took a spear in the guts this week alone. The yukfoos have some newfangled spear-thrower thingie that works way better than bare hands. If we keep this up sooner or later we're going to run out of foot soldiers.

[Tribal Leader:] Nonsense, my friend! The yukfoos are pure evil! If we don't fight them to the last man they'll steal our women! They'll steal our food! They'll slit our throats! They'll destroy everything we stand for! Get out there and kill! Kill! KILL!

[Foot Soldier:] Yeah, I keep forgetting! Arooo! Ngrglar! [Runs off waving spear.]

[Tribal Leader:] Damn, that was close. [Turns.] Hey, Jeeves, run down to the village and drag me up a woman, willya? See if you can find one I haven't had in awhile. If her husband objects, just slit his throat. Oh, and if anybody down there has any meat, grab it while you're at it. Cut off a chunk for yourself if you want, but bring me as much as you can. Man, I haven't eaten for an hour and a *half!*

[Exeunt omnes.]

There's nothing worse than tribal foot soldiers who begin to think about their situation in the abstract. They might just quit the game, run off and start a new tribe somewhere else, or possibly sneak back with one of the yukfoos' spear-throwers and nail the tribal leader through an eye socket. This would bode poorly for the continuing success of the tribal mechanism. So the blind watchmaker tries lots of things, and what works is a way to amplify tribal loyalties and cloud the emerging rational mind. This new countermeasure is anger.

From my readings in ethology and anthropology, it seems like anger is a fairly recent tool in the kit compared to the tribal mechanism. Animals seem blase about killing, as do most of the newly contacted primitive tribes that Jared Diamond studied decades ago. It's very much a "nothing personal, Mac" kind of thing. What we call psychopaths may simply be throwbacks. They don't get angry. They don't even get worked up. When they feel so moved, they don't think about it. They just kill. The possibility that they themselves may die in the attempt doesn't seem to bother them.

Anger makes it possible to bypass abstract thinking in ordinary people and make them do stupid and damaging things, ideally directed against other tribes. It can be triggered by a number of things. Still, nothing seems to rev it like the notion of Us vs. Them.

Killing members of other tribes is now illegal in the developed world, but tribal leaders still stoke the fires of tribal anger to keep their omegas outward-facing and loyal, and damaging opposing tribes whenever possible, through the ballot box if not through the eye sockets. The end result is that the tribal mechanism remains very much alive, and very much at work transferring wealth and sexual opportunity up the turtle pile to tribal leaders at the top. Why anybody plays the game is a puzzle, unless it really is genetic and those who do it simply can't help it.

Note well that this is a hypothesis. I'm not a sociologist, psychologist, or anthropologist, and I have no idea how one proves such things. I'm guessing it can't be proven at all. But man, that's how it looks from my window.

Anger Kills

From Jeff Duntemann's Contrapositive Diary, November 16, 2016

Anger literally killed my grandfather. I mean literally literally here, not figuratively: My grandfather Harry G. Duntemann got furiously angry, and he died. This is one reason I've tried all my life to be good-natured and upbeat, and not let piddly shit (a wonderful term I learned from my father) get me worked up. This worked better some times than others. (Once it almost didn't work at all. I'll get to that.) Practice does help. However, in the wake of the election, a lot of people whose friendship I value are making themselves violently angry over something that may be unfortunate but can't be changed. This is a bad idea. It could kill you.

Consider Harry Duntemann 1892-1956. He was a banker, fastidious and careful, with a tidy bungalow on Chicago's North Side, a wife he loved, and two kids. One was a model child. The other was my father. Both he and his son were veterans of the World Wars, which is one reason I mention them today. My grandfather, in fact, won a medal for capturing two German soldiers in France all by himself, by faking the sounds of several men on patrol and demanding that they come out with their hands up. They did. He played them good and proper, and nobody got hurt.

He had an anger problem. Things bothered him when they didn't go his way. Family legend (which I've mentioned here before) holds that my father comprised most of the things that didn't go his way. His anger isn't completely inexplicable. Harry worked in a bank, and was for a time the chief teller at the First National Bank of Chicago. You don't get to do jobs like that if you're sloppy, and if you spot errors, you track them down like rats and kill them.

Harry was the sort of man who really shouldn't retire, but retire he did, at age 62. He bought a lot in tony Sauganash and had a fancy new house built. I honestly don't know what he did with his time. He golfed, and taught me how to do simple things with tools when I was barely four. He worked in his garden and his vegetable patch. My guess: He was bored, and what might not have bothered him when he oversaw the teller line at Chicago's biggest bank now preyed on his mostly idle mind.

One day in August 1956 a couple of neighborhood punks vandalized his almost-new garage, and he caught them in the act. He yelled at them, and they mocked him. He yelled more. They mocked more. Finally he just turned around, marched into his house, sat down in his big easy chair…

…and died.

He was healthy, a lifetime nonsmoker, trim, not diabetic, and not much of a drinker. I suspect he was more active in retirement than he had been during his working life. He had no history of heart disease. He had no history of *anything*. Anything, that is, but anger.

I ignited a smallish firestorm on Facebook yesterday when I exhorted people who were angry over the election to just let it go. Most of them seemed to think that "letting it go" meant "accepting it" or even condoning it. Maybe in some circles it does. I don't know. To me it means something else entirely, something that may well have saved my life.

As my long-time readers know, I lost my publishing company in 2002. It didn't die a natural death. I can't tell you more than that for various reasons, but Keith and I didn't see it coming, and it hit us hard. I put on a brave face and did my best. Once I was home all day, though, it just ate at me. I was soon unable to sleep, to the point that I was beginning to hallucinate. To say I was angry doesn't capture it. Depression is anger turned inward, and I became depressed.

I had a lot of conversations with Bishop Elijah of the Old Catholic Church of San Francisco. He was getting worried about me, and in late 2002 he Fedexed me a little stock of consecrated oil, and told me quite sternly to anoint myself. I did. (After I did, I laughed. Would Jesus haved used FedX? Of course He would. Jesus used what He had on hand to do the job He had to do. Catholicism is sacramental, but also practical.) Elijah diagnosed me pretty accurately when he said: *You're hoping for a better yesterday. You won't get it. Let it go.*

It took awhile. It took longer, in fact, than Bishop Elijah had left on this Earth, and I struggled with it for years after he died in 2005. The company wasn't piddly shit. It was the finest thing I had ever done. How could I let it go?

I thought of my grandfather Harry every so often. And eventually it hit me: Those little snots didn't kill him, as I had thought all my life. They played him, and he killed himself with his own anger. "Letting it go" cooked down to protecting myself from myself. I'll never get my company back, but I can now see it from enough of a height to keep my emotional mind from dominating the memory. I learned a lot as a publisher. I made friends, and money, and reputation. I supervised the creation of a lot of damned fine books, and won awards. Losing it was bad, but life around me was good. (Carol especially.) I could choose to obsess, and probably die before my time, or I could recognize the damage my anger could do and turn the other way. I'm not sure how better to describe it. It was a deliberate shift of emotional attention from my loss to new challenges.

This isn't just a theory of mine. Anger kills by keeping the body awash in cortisol, which causes inflammation of the arteries. The inflammation causes loose lipids to collect in arterial plaques, which eventually block an artery and cause an infarction. Plug the wrong artery at the wrong time, and you're *over*.

Anger is a swindle. It doesn't matter if it's "righteous anger," whateverthehell *that* is. Anger promises the vindication of frustration and disappointment, and delivers misery and early death. When I've seen people online turning bright purple with fury the last couple of days, that's what I see: Good people being played by the desire for a better yesterday. It won't kill most of them. It may well kill a few. It will lose them friends. It will make other people avoid them. It may prompt them to eat and drink too much. It is basically making them miserable, to no benefit whatsoever.

When I say "let it go" these days, I mean what I said above: *Protect yourself from yourself.* Call a truce between the two warring hemispheres of your brain. Turn to something else, something you can change, something that may earn out the effort you put into it with knowledge, skill, and accomplishment.

Believe me on this one: There is no better yesterday. Don't go down that road.

You may never come back.

30 Lessons I've Learned in 61 Years

From *Jeff Duntemann's Contrapositive Diary*, November 17, 2013

1. Defy convention.

2. Question authority.

3. Keep your promises.

4. Nothing is simple. Simplicity is bait on somebody else's hook.

5. Never wear anybody's advertising but your own.

6. When you think you've heard too much Gustav Holst, play some Madonna.

7. Friends are a revenue center. Enemies are a cost center.

8. Never believe anything an angry person says, especially when they're not angry.

9. Fat makes you thin. Sugar will kill you.

10. Political parties exist to take everything you have and hand it to psychopaths on a silver platter.

11. Fathers *matter*.

12. Time shatters what cannot hold, and perfects what cannot be broken.

13. If you can still wear a shirt thirty years later, you know you're doing OK. This is a good reason to keep a shirt or two for thirty years.

14. Join a political party and you're selling youself into slavery.

15. *Evil* is the root of all evil. There is no middleman.

16. Love matters way more than who's got a plug and who's got a socket.

17. Don't try to make a bowling ball out of 2 X 4's.

18. You're not really an adult until you can run around the house in your underwear, reciting Dr. Seuss at the top of your lungs.

19. Certainty is a species of mental illness.

20. Self-esteem is confidence without calibration.

21. Think outside the box. Then make something out of the box.

22. If you see a piñata, remember that somewhere close by is a blindfolded person swinging a stick.

23. Pitch can be useful. It's *politics* that defileth all it toucheth.

24. Don't settle for an iron will. Gram for gram, aluminum is stronger.

25. A dog is a fingertip of the Almighty, thrust briefly into our lives to measure the breadth and depth of our kindness. Remember Whose fingertip it is.

26. Dance, especially if you're not good at it.

27. Stand by your spouse *no matter what.*

28. They build too low, who build beneath the stars.

29. Kick ass. Just don't miss.

30. *Think!*

Getting Past Nagasaki

From Jeff Duntemann's Contrapositive Diary, August 7, 2005

We're now approaching the 60th anniversary of the end of World War II. I have something odd and upbeat to post on VJ-Day, assuming I can find the files. If not, I have some scanning and OCRing to do again, sigh.

Sigh, indeed. Yesterday was the 60th anniversary of our dropping a nuclear weapon on Hiroshima. Many or even most people who are not completely ignorant of the history of WWII or totally wigged out by nuclear weapons understand the necessity of Hiroshima. The world stood stunned as the smoke cleared, and against a threat like that, Imperial Japan would have caved in days. Then there was August 9. Why did we have to do it again?

First of all, avoid the temptation to second guess and judge the people who lived the era and bore the responsibility. People were dying across the world, not by hundreds or thousands, but by *millions*. Whole nations and peoples were virtually wiped off the planet. How well would *you* have handled it?

I've been boning up on my 20th century history lately, through several books like *The Great Influenza*, *The Fall of the Dynasties*, and *The War Against the Weak*, along with a quick flip through the marvelous 1966 *American Heritage Picture History of WWII*, though I wept when I read my father's notes in the margins. Good God, he was *there*, in the thick of all that hell, dust, and death. He, at least, got back alive, as a man named Robert Williams, who might otherwise have been my father, did not.

I think I understand Nagasaki. I don't like the understanding I have, but I understand: WWI ended scarcely twenty years before WWII began. The death-stink of Verdun remained vivid in the memories of those who survived it. (They are *still* digging unexploded ordnance from those now-peaceful fields!) The world seemed to be recognizing a pattern: Every generation, a strange psychosis reached some sort of critical mass, and erupted in increasingly deadly conflicts between nation-states that (by 1945) should long have known better. Even as Nazi Germany collapsed, I think that forward-looking people were charting the line between 1870, 1914, and 1939, and did not like the shadow they saw ahead. The points were growing closer, and the death toll higher, each time that the world went to war. Patton knew what Stalin was, and although he was forbidden his plan to take Moscow, I think his superiors eventually came to understand Patton's insight. I'm almost certain that the next European war would have come by 1955, and a nuclear-powered Soviet Union would have reduced much of Europe to sizzling ash.

Instead, we took Nagasaki. One might have been a fluke, or good luck. Two in four days was a statement that could not be ignored. In a sense, the American leadership was telling the rest of the world, Stalin and every other emerging nationalist psychopath who might be watching: This..nonsense..will..stop...*now*.

I mourn for Nagasaki, as I mourn for the Jews, and the Russians, and the Ukraine, and my mother's high-school sweetheart. It's been quiet now for sixty years. There has never been another nuclear attack. In my view, there has never actually been another war. (Those who consider Iraq I or II or even Vietnam a "war" need to read more history.) The world turned a corner in 1945. We stopped connecting the dots, and there is some hope that the horrible line between 1870, 1914, and 1939 will not be drawn again. 75,000 people died at Nagasaki, but had they not died, 100,000,000 would almost certainly have perished the next time the world erupted.

Remember: *There is no such thing as pacifism.* Doing nothing is doing something. There is no escaping responsibility. There are no good choices. All we can do is bless our dead for what their lives have purchased, and move on.

Happy Beginnings vs. Happy Endings

From *Jeff Duntemann's Contrapositive Diary*, June 17, 2009

Sixty years ago today, my parents were married, at St. Mary of Perpetual Help church on West 32nd Street in Chicago. It was a remarkable event, not so much because history will consider my parents remarkable (though I do) but because it was, well, unlikely. This remarkableness was not unique, but occurred countless times across America in that era, as social and ethnic barriers that had stood for centuries started to crumble, and men and women began to marry for love and not to satisfy family demands.

Consider Frank William Duntemann, the only son of a bank officer at the First National Bank of Chicago. He had been born and raised solidly middle class in East Rogers Park, of a German father and an Irish mother. Hard-headed, ironic, optimistic, stubborn, bright, slightly snotty, and short–5'6" of solid muscle, fearless and (especially as a young man) a little pugnacious. He drove his parents crazy sometimes, running off to join the Army in 1938 when he was only 16 (the Army sent him home) and getting suspended from Lane Tech for beating the crap out of the six-foot-tall president of the Lane Tech Nazi Society, after the Nazi had made the *serious* mistake of stabbing my father in the stomach with a wood chisel during a shop-class argument.

And consider Victoria Albina Pryes, the youngest of ten children, born of penniless Polish immigrants in a ramshackle farmhouse in Stanley, Wisconsin. Artistic, fretful, possessed of a beautiful voice, pious to the point of mysticism, and ethereally beautiful, she trained as a nurse in Chicago after WWII and struggled with the question of what to do with her life. Her family thought she should become a nun, because her high-school sweetheart had died in the War, and that could only be a Sign. But she held back, and one day in 1946 a nursing school friend suggested a double date. Mary's boyfriend knew this interesting guy from the North Side…

Frank was smitten. Victoria was terrified. He asked for her phone number, and in a panic she made something up. Undeterred, the man who had slept through the bombardment of Monte Cassino sent a postcard to her nursing school (we have that postcard) asking her to get in touch. Even though torn between what she felt to be her family and religious obligations and her own infatuation, she did. Not sure what to expect from a man so far removed from her ethnic heritage and socioeconomic class, what she found was passionate friendship. In 1948 he asked her to marry him. By then, there was no hesitation.

But it was not without challenges. Frank's parents were furious. They had expected him to marry a nice German girl from the neighborhood. Instead, he had chosen a Polock farm girl living in what they considered the slums. Harry Duntemann was not a man to be trifled with, and he told his son to break it off. Harry had managed to browbeat Frank into a bookkeeper's job that he hated, and was nagging him to return to Northwestern for a degree in business. But the War had changed Frank, as it had changed thousands of men who had been frightened boys the day after Pearl Harbor. Frank took his father aside and told him, "Look, I've made my decision and it's not open to discussion. I'm going to marry Victoria, and then I'm going to Georgia Tech to get my engineering degree on the GI Bill. If you want us to come back here, and if you want to see your grandchildren, you'd better start seeing things my way."

Harry, perhaps recognizing his own stubbornness in his son, gulped and agreed. (And to ensure that his son *would* return from Georgia, helped buy him a house–on the North Side.) And so on that gorgeous June day in 1949, my parents made their Happy Beginning, bridging two widely disparate cultures, he confidently, she (as always) apprehensively.

By any measure it was a successful marriage. Frank and Victoria changed one another: He taught her confidence, and persuaded her that she was beautiful and worthy; she taught him moderation and compromise. She was not sure she wanted children, but he did; he was not sure that a gentle style of childrearing would work, but she did. They were in fact spectacular parents. They read to us, they bought us books, they insisted that we speak correctly and tell the stories of our days at the dinner table. My father threatened to call the Alderman if the Chicago Public Library refused to give me a library card for being underage. (I was six; you had to be seven.) I got the card. He gave me money for electronic parts and bought me a microscope; later, when I was deeply into junkbox telescopes, my mother always had a dollar for *one more* pipe fitting. We were not especially flush, and were taught frugality, but money was always there for things that mattered. Stubborn as he was, my father had the courage to avoid his own father's mistakes: He told my sister and me that no matter what careers we chose, he would support us in that choice.

My father loved my mother fiercely, and the lesson was not lost on me. More than once, when I was sitting at the kitchen table doing my homework, my father came home from work a little early, went up behind my mother at the stove, kissed the top of her head, and told her he loved her. When I was fifteen, he made it explicit: "Love comes out of friendship. If you're lucky and smart, you'll marry your best friend." I did as he said (and also as he did) and no better advice has ever been given to me.

Happy beginnings are often easy. Alas, happy endings are not automatic. I've told most of the rest of the story here before. In 1968 my father was diagnosed with oral cancer, from his two-pack-a-day habit he had picked up in Italy during the War. He fought back, and it took nine years, but the cancer killed him a piece at a time, in a gruesome progression that still gives me nightmares. It broke his spirit and finally took his mind; at our wedding in 1976 he was weak and confused. By 1977 he no longer knew who I was, which broke my heart, and in January 1978 it was finally over.

My mother was never the same. Living alone in their house for another 18 years allowed her to brood on questions of divine justice that had always haunted her. What had she done to offend God? How had she failed? My mother's understanding of Catholicism was suffused with peasant superstition amplified to absurdity by her odd mystical personality. It was a cruel and often bizarre religion, full of prophecies and portents and dark powers, overlaid against the looming background of an angry God and an animate Hell. She was tormented by hideous dreams of accusing demons, dreams that may have led (as Gretchen and I have speculated) to the insomnia that plagued her last years. She was literally afraid to sleep, fearing what she might dream. Her doctors tried various drugs, but nothing helped, and even with Gretchen and Bill's constant companionship and loving care, she lost her ability to speak, and slowly withered away to almost nothing. When I carried her out to Gretchen's van the day before she died, she may have weighed fifty or sixty pounds, and looked like a victim of the Biafra famine.

It made me furious then, and I still get a little nuts thinking about it. How can people who tried so hard, who loved one another so truly and unfailingly, who were generous and industrious and offered their children nothing but unconditional love, suffer such hideous ends? *Where's the justice here?* The answers are complex, if in fact they are answers at all, and my readings on theodicy have been scant comfort.

Yes, they deserved a happy ending. And because they never had that happy ending, the day after my mother died in 2000, I sat down and wrote them one. [See p. 158, "Homecoming."] They allowed me to be a writer, which is not as secure a career as an engineer (or almost anything else) so it was the least that I could do.

Still, the question stands: Where is the justice? If God does in fact exist, He owes me an answer to that painful question–but if God does in fact exist, (as I think He does) He's already provided the answer–and the happy ending–to those, like my parents, who are farther along the Great Path than you or I.

Homecoming

Published on the Web, June 17, 2009

End of the line, M'am," said the bus driver, grinning over his shoulder as he pulled the old bus into a tight right turn.

She shook her head slightly from side to side, feeling herself awaken from troubled sleep. She smiled and tried to thank him, but for a long moment, in confusion, she couldn't find any words at all. When had she gotten on the bus?

The driver stopped the bus with a little lurch and left his squeaky brown leather seat, to take her arm as she stood, puzzled, in the aisle between the rows of empty seats. He helped her forward, she on wobbly legs and strange heels on shoes that looked at once strange and familiar, like something out of a warm old memory. She hadn't worn shoes like that since, what...1949?

"Easy, M'am," the driver said as she stumbled again, and with such tenderness that she had to turn and look at him. He was ancient and yet had the jolly eyes of a boy, framed by tight grey curls over his ears.

"I..I was asleep," she said, and it felt good to speak again. "Thank you." She felt her strength coming back, and looked at the twin line of ad placards above the bus windows, pitching Wrigley's Gum and Yoo Hoo soda, looking older and out of a simpler time and place, a time and place where there wasn't so much pain, nor so much hopelessness.

The driver stood by the front door of the old bus, and held her arm in his strong, dark hand while she took the two steps down and the third out into the wet summer night.

Moments later, the bus pulled away, and she was alone, with the night and the streetlamps, and off through the drizzle warm yellow light shone from the townhomes along the avenue. The air had a faint sweetness in and about it, something she remembered and couldn't quite place. She spent a long moment struggling to remember, then abruptly wondered where she was. She was ill, she was...old. Mostly (as for many years now) she was alone.

Then she looked again, and he was there.

"Hi, Babe." The nearby streetlamp glinted like the evening star on his rimless crystal glasses. "Been waitin' for you."

"I..."

"Shush." He was face to face with her now, in an old Eisenhower jacket that had seen some hard use.

When nothing else seemed especially real, why hold back? She threw herself into his arms, sobbing uncontrollably, feeling his strong hands gently stroking her back, feeling the warmth that she recalled so well, the commitment and certainty she had always felt near him, the friendship that enlivened her, and that old fire beneath it. With a roar it came up within her, unbidden and not yet believed: *I am no longer alone.* "Oh God, my Lord, Jesus, Mary, and Joseph..."

She looked up into his face, now streaked with tears as well, but *young*, strong, whole, grinning indomitably as he had the day he had lifted her veil to make his promises of forever. She kissed him quickly on tiptoes, like the young woman she had been when she had last worn these shoes and stood amidst these trees on this ancient southern street. "So I'm dead too?"

He shook his head. "No. You're home. *There is no death.* And supper's waiting. C'mon. There are some friends I want you to meet."

There was no disbelieving the feel of his hand surrounding hers. She grasped his arm and without thinking or caring which way was which, she started off in step with him. For blocks they walked, while with each breath she felt old strength coming back. He said nothing, but she felt his presence everywhere around them, hovering like a bright cloud that kept the slow rain away.

Then, in front of them, a strange hiss and the ugly tang of rotting trash smoldering in an incinerator. She looked up, to see a thing like a rat the size of a dog rearing on its hind legs, fixing them with burning red eyes and pointing at her with a clawed hand.

He stopped, and giving her arm a squeeze, stepped calmly a half pace forward.

He said nothing, but a change came over him in a heartbeat. From his face welled up a light, bright but not burning, not hot but golden in its *rightness*, unlike any light she had ever seen. He raised his hands, not in anger but in forgiveness, or (perhaps) dismissal, and the holy light was shining from them as

159

well, his face transfigured beyond humanity to something greater, something that lived in the Light, and bore that Light like a sword to places where there was still darkness.

The dark creature howled, inhaled, and shrieked a horrid sound, transfixed in the Light until even its twisted bones were laid bare. It turned and bolted, whimpering in pain, until it leapt down a drain in the gutter. The rain bore away its stench of decay and heavy smoke.

As quickly as the Light had come, the Light faded—and it was he again as she had known him so long ago.

He turned back to her, unsure what to say, almost embarrassed. His words were close to a whisper: "I have looked upon the Face of God. Nothing evil can ever approach me again." There was something all around him that she hadn't quite noticed before, something that her eyes didn't quite see but that her heart could never again ignore. It recalled something of that golden light that spoke of ineffable power from far beyond her, and she knew that while that power encircled her, she would be safe from all threats, seen or unseen.

For blocks they walked in silence, seeing warm lights in windows, hearing indistinctly the sounds of other people speaking, singing...and far away, dancing? Without her quite seeing where he'd come from, a little blonde cocker spaniel was now walking beside them, tongue lolling as its nails made a happy little sound on the sidewalk. The bus passed beside them with a diesel sigh, and from within the old brown driver waved.

She shook her head and laughed a little, squeezing his strong right arm and leaning her head over to touch her cheek to his shoulder. "So this is heaven? They could have given you a better jacket."

He shrugged. "Hey, I haven't worn through the pockets yet. Got a little life left in it. This isn't heaven, either."

She stiffened a little. Just like him, to argue such an inarguable point. The warm drizzle gave it the lie. If he were here, how could it not be...

"I guess you'd call it purgatory."

"Purgatory! But I'm not suffering...I mean, you're here, and it's warm and clean and I can smell the trees in bloom!"

He shook his head, smiling. "Purgatory is a place for purging what you can't take with you into heaven. This can mean sadness as well as sin. When you're sad for so many years, it leaves a mark on you. This is a place where you get what you should have had, where we can have the years together that we didn't, where little by little you'll leave behind the weariness that sadness soaks your bones in. Then one of these days you'll realize that you're somewhere else entirely, someplace I've been but couldn't even begin to tell you about."

Lost in the oddness of it all, she realized that she was standing in front of a familiar old door. Not luxurious, but...it would do, it would do. *Do?* Suddenly she understood that there was justice as surely as there was pain, and with that thought the heaviness began to fall away.

"Don't even try," she said softly, as the brass knob turned in his strong hand. "I want those years first."

"You got 'em!" he said, and with a whoop he swept her up bodily off the porch, and carried her triumphantly through the cast-wide door into eternal light.

PART 3. PARODY

Humor is hard. Trust me on that. It may be genetic; my father could be very funny, and my sister and I both have a knack for humor. My knack surfaced early in a weird subsubsubgenre of humor: Song parodies. In Catholic grade school we sang songs from little songbooks published by the Sisters of Providence, who taught at our school. I wrote different words for some of the songs, often making fun of the (abundant) quirks of our teachers. Years later I discovered that there was a whole separate term for song parodies: "filk songs," which came out of the science fiction and fantasy fan community. I think it may have originally been a portmanteau of "filthy folk songs," though I've only encountered a handful that were about sex. Crazy world, and rarely crazier than in SF fandom.

Once I discovered SF fandom at the Clarion SF writers' conference in the summer 1973, I was all-in on the filk concept. I wrote parodies of "The Cassions Go Rolling Along," "The Beer-Barrel Polka," "The Marine Hymn," and several songs from, of all things, good grief, "West Side Story." All of them had an SF or fantasy theme.

I didn't start writing poetry parodies until I was editing *PC Techniques*, and had a "last page" dedicated to humor, crazy ideas, and none-of-the-above. In truth I'm not much of a poet. What original poems I've written were mostly pastiche of poets I've enjoyed reading, but they were pastiche (i.e., written in those other poets' styles) and not parody.

In this section I've collected the better-known of my parodies, of poems, songs, and in one case, a column in another magazine, *PC Week*.

Ten Megs of RAM

From *PC Techniques Magazine* #40, October/November 1996

I cram RAM.
I cram RAM.
RAM-I-Cram.

I'm RAM-I-Cram and I sell RAM.
Do you have ten megs of RAM?

I do not have ten megs of RAM.
I do not need them, RAM-I-Cram.

Do you code in C++?
Simulate a railroad truss?

I never code in C++
Nor simulate a railroad truss.
I do not have ten megs of RAM;
I do not want them, RAM-I-Cram!

Play a game of Network Quake?
Or VR Vampire! Here's your stake!

That is not my idea of fun;
I only like my steaks well-done.
I do not need ten megs of RAM;
I will not buy them, RAM-I-Cram!

Build a swapfile like NT's!
Multitask your EXEs!

RAM, you can nag till hell might freeze;
Up in the air or on your knees:
I do not want ten megs of RAM;
I will not use them, RAM-I-Cram!

Fill your laptop to the brim!
Take them! Take them! Here's the SIMMs!

Not in my lap! Not COD!
It's all a trap! RAM, let me be!
I do not want ten megs of RAM;
I can't afford them, RAM-I-Cram!

Hmmmm...have you *tried* ten megs of RAM?

(Long pause, mouse clicks, disk noises)

You've won again, you nasty man;
I must have those ten megs of RAM.
My spreadsheets load in half a flash,
(Although you've drained my bank of cash.)

Well, sure, the payoff's worth *that* risk;
Now...

...do you have four gigs of disk?

Staring at Procs, in Maintenance Mode

From *PC Techniques* Magazine #18, February/March 1993

(With Apologies to Robert Frost)

Whose procs these are I'm sure I know
He lives up in the mountains, though;
He would not care to help me here—
He sold his options long ago.

My CS prof would think it queer
To pause without a breakpoint near
Down where our hackers never go—
(At least not since that mess last year...)

Now, why XOR before INT0?
To clear ES on overflow?
Why doesn't that corrupt the heap?
I'm sure that I'll just never know.

His code is subtle, fast, and deep.
But I have schedules to keep,
And bugs to fix before I sleep;
And bugs to fix before I sleep.

Radio WONK

From *PC Techniques* Magazine #9, August/September 1991

Gooooooooood Morning Hackertown! It's 0600 and time for all you wonks and wonkettes to cold boot and hit the road to byte the books and blast those bits and beat the bugs out of this brand new smokin' rockin' executin' Monday morning! (*Gong, rooster crowing*). Now.....Big Billy Gigabyte Gates, your Hackertown morning man, (cheers, wolf howls) knows that there was a little bit too much boppin' in the belly of the BIOS this weekend, so just throw some cold water into those bleary eyes and climb into your best bluejeans and worst sneakers, but make sure you have your shorts on right side out! (*Gong, bird noises, sound of flushing toilet*).

And to kick off the Monday Makefile, (*shotgun blast*) let's crank up the megahertz to raise the daytime flag and pick up a downsouth Cajun hackabilly oldie from the dawn of time, Adam Osborne and the Kay-Pros bashing out "Jumbled I/O:"

[Author's hint: "Jambalaya," with a bundle apology to Hank Williams...]

Diablo, she gonna go, me-oh-my-oh;

She gonna take paint off the wall, go sky-high-oh;

Hazeltine, sure sound mean, me-oh-my-oh;

Sunuvugun, gonna have big fun in the BIOS.

Chorus:

Jumbled I/O, me-oh-my, what the hell-oh,

'Cause tonight I'm a-gonna make my machine go;

Off she run, like a gun, then she die-oh;

Sunuvugun, gonna have big fun, in the BIOS!

Me got a bone, pick on the phone, call Intel-oh,

'Cause when me write newline to LST she ring the bell-oh;

PIP to B:, FORMATS C:, then me cry-oh;

Sunuvugun, gonna have big fun, in the BIOS.

Chorus

Plotter drum, she start to hum, disks is buzzin';

Error come up on the console by the dozen:

Bad sec-tor: accept, ignore, or retry-oh;

Sunuvugun, gonna have big fun, in the BIOS!

Final Chorus

Man, they don't hack 'em like that anymore! That tune hit #1 on the Radio WONK Bit Parade for 175 consecutive machine cycles back in 1978, and it still spins as good as an 8" floppy disk. (*Creaking door*). And speaking of the Bit Parade, don't forget that today is a freshly allocated week, and all you wonks and wonkettes can pick up your brand new absolutely free Radio WONK Bit Parade Top 10H Survey Slips down at your neighborhood Bits'n'Hits stores first thing this afternoon, so go get 'em! (*Cheering, cannon blast*).

And before I free memory and call the destructor on this morning's show, Big Billy Gigabyte Gates himself has a request. I once heard somebody say that there is more to life than hex, bugs, and rock and roll, and baby, if you ever figure out what, then you just call the toll-free hack support line at 1-800-ASM-CODE and tell me, 'cause I DO NOT BE-LIEEEEEEEEEEEEEVE IT! (*Gong, siren, sound of breaking glass, crackle of laser printer jam, pop of surge protector...*)

THE RADIO WONK BIT PARADE TOP 10H!

01H: Girls Just Wanna Type RUN—Cindy Looper

02H: I've Been Through Some Changes On A Clone With No Name—Taiwan

03H: Load and Go—Emerson, Lake, and Pascal

   Don't It Break My CPU

04H: TWO SIDED WINNER!—Crystal Fayle

   Why Have You ACKed the One You NAKed Me For?

06H: A Bar Named Foo—Johnny Cache

07H: Who's Hackin' LISP On Your Old AT

      (While You Was Out Hackin' LISP?)—James Brain

08H: Like A Version—Mañana

09H: Leanin' On a Backplane—Adder, Poll, and Carry

OAH: Multibus—The WHILE

OBH: The Man Who Got Every Tree Balanced—C. Bitme

OCH: The NO-OP Dance—Pointer Sisters

ODH: Breakin' Up With OS/2—Neil's Ahackah

OEH: There's No Parity and I'll Trash What I Want To—Leslie XOR

OFH: Al-Go-Rhythm—The Tokens

10H: Your Mama Don't Strobe, and Your Daddy Don't Loop and Poll

      —Login & I'llseeya

WONK—BIG 86 ON YOUR RADIO DIAL!

Brought to you by Bits'n'Hits—The Biggest and Best Books, Bytes, & Be-Bop Shop in Greater Hackertown or anywhere else in southern Ohio where people still live!

Wagger Central

From *PC Techniques Magazine #8*, June/July 1991

As 110° plus temps and searingly sunny summer skies started scorching Scottsdale, Mr. Byte climbed into his climate-controlled waterbed-equipped Sky Kennel and jetted off to New York City in the belly of a 747, licking his chops at the thought of all the morsels lying under the tables at PC Expo.

Fuzzy logic? Old sock, yapped the Yipper. The latest thing out of top-secret Japanese research labs is Artificial Ambivalence, a new technology created for MIS/DP analysts who simply can't decide how they feel about something. Real hot, buzzed the Byter, who saw Esther Dyson and Ruthann Quindlen eyeball-to-eyeball at a refreshments table on the show's periphery, toe-wrestling over who would have a lock on abusing the concept in print.

The Dirt-Digging Doggie dared dodge delirious dancers at the drink, dine, disco and demo party thrown by Lotus, where top-level managers were looking at and feeling everyone in sight, alert for hidden cameras and bugs. Ha! huffed the hirsute howler. You want bugs, look behind my ears—or inside the new Lotus four-dimensional spreadsheet, code-named Fourflusher. The product was designed to augment productivity by performing all of its recalculation in the future, thereby providing instantaneous response on even the hairiest spreadsheet. But by lying doggo in some influential laps, Byte was able to sniff out the real poop: Fourflusher often returns values based on data that the user hasn't typed in yet.

Time to pull on your black net stockings and do the Timewarp, opines the Bemused Bichon. The Master of Mange followed the smell of seafood and several savvy alley cats to IBM's hush-hush research lab on Wakaholinit Island, a silver spoon's toss from Nantucket, where IBM's researchers are burning the midnight chafing dish creating a new species of processor called BISC—the Biological Instruction Set Computer, created by installing the complete nervous system of a lobster in a 466-pin DIP. According to papers hidden under a 20-gallon pot in the lab, the BISC chip was created to elegantly solve certain classes of problem that plague the Massachusetts Old Money crowd, like what wine to serve with red fish, and how long to sauté truffles before serving them. Mr.

Byte's furtive inspection of IBM's project documents revealed many problems yet unsolved—like what to do with the lobsters after removal of their nervous systems. And why did lobsters get the nod for the project? As an unnamed IBM researcher wrote in the secret log, "We needed something smarter than your average Kennedy—the lobster was an obvious choice."

Is there a Turbo APL in your future? Do banana slugs make good banana daiquiris? (Only with salt, cautions the Cur.) If not, then why did the Whiskered Woofer whiff all that Greek cooking down at Borland's Scotts Valley HQ? And what of that book entitled *Reshape Yourself* spotted half-hidden amidst the papers on CEO Kahn's desk? And the memo fragment retrieved from a dumpster (along with some savory scraps of said Greek cuisine) instructing a staffer to "... roll a deal with that guy from Matrix..."? Pretty obvious to me, beamed Byte.

Then it was up the Left Coast to Redmond, where the Undercover Un-Poodle unearthed an utterly unbelieveable plan to buy Microsoft. While lifting his leg in the loo, Mr. Byte listened to a local loonie brag that he intended to put the entire megacorporation on his credit card and then tell the bank it was never delivered. Just might work, mused the Midnight Mongrel, except that the OS/2 V2.0 documentation set would never fit in a UPS truck.

The PC magazine business is as healthy as ever, crooned the cocky canine, but there may be truth to tidings that computer publishing giant Ziff-Davis was preparing to can several key employees. Mr. Byte relates how on his last trip to Boston he spied a grey cat in a business suit yowling down Boylston street with a box of nightshirts under one arm, a string of tin cans tied to his tail.

Heard any good shaggy dog stories lately? Toss a juicy bone to Mr. Byte and get a very exclusive rhinestone collar signed by the Peripatetic Pup himself. Don't keep him panting ....dump your load at 1-900-POO-SCOOP.

I have a sneaking suspicion that a lot of readers won't quite understand this one. Some explanation may be in order. In the wake of its publication in the magazine in 1991, I got paper letters (email was still uncommon) saying something like, "I'm sure this would be hilarious, Jeff, if I could only figure out what you were making fun of."

Touché. That's the problem with parody, and I've had the problem more than once: Writing a parody of something few people have experienced almost never works. What I was making fun of here was a marvelously droll column in *PC Week* called "Rumor Central," bylined Spencer F. Katt but gang-written by several *PC Week* staffers. It was the thing most people read first when their copy of *PC Week* arrived in the mail. It was a sort of PC technology gossip column written by an anthropomorphic cat in a nightshirt.

So what's wrong with that? Just this: *PC Week* was a "controlled-circulation" publication. This meant that it was sent free to anyone who could qualify for a subscription by being a technology vendor or a "communicator" (PR reps, ad reps, magazine staff, etc) within the IBM PC industry. It didn't go to a lot of people, but the people it went to were precisely the kind of people who bought (generally big and expensive) ads, or who shaped opinion about PC hardware and software. I'm not even sure you could purchase a subscription for money. So although a handful of my magazine's subscribers might have been eligible, most had no idea that a weekly industry tabloid called *PC Week* even existed.

The other thing modern readers may not understand is who Mr. Byte is. Mr. Byte 1980-1994 was my first dog as an adult, a bichon frise whom I mentioned in all my books and here and there in the magazine. My regular readers all knew who he was. It was "Rumor Central" they'd never heard of.

My suggestion here if you're interested at all is to google "PC Week Rumor Central" and read a couple of the columns from back in the day. Then reread "Wagger Central" with Spencer F. Katt fresh in mind. Yes, you'll laugh. And yes, I learned my lesson.

The Lovesong of J. Random Hacker

Originally published on the Web

(With apologies to T. S. Eliot)

(Doctah Kurtz, he dead. A GOTO for the old guy.)

Let us go then, you and I,

For fast Chinese and talk of years gone by

Filled with random jumps and custom cable;

Let us go, recalling joys of FORTH and MUMPS,

The cluttering lumps

Of threaded code in frantic ten-hour hacks

To get that midterm project off our backs:

With code that twisted, doubled-back and bent

And set into cement

But came through with an underwhelming "B"...

Oh, do not ask, "What was it?"

I don't care what it does, just how it does it.

On the Net the expert systems come and go,

Bragging about how much they know.

Over yellow chad that chattered out from teletype machines,

Over yellow tape that rattled out encoding fever dreams

That curled into the data center trash;

We lingered, inventing novel sort/merge schemes,

Or ways to thwart collisions when we hash--

And seeing that we'd been logged in since late last week

Took one last slug of Jolt and fell asleep.

On the Net the expert systems come and go,
Bragging about how much they know.

No! I am not Bill Gates, nor would I want to be;
I'd rather parse the fish than own the knife;
(Imagine! Having moby bux but chained
to ninety million lusers, what a life...)
Am a flamer, goateed, pallid, overweight,
Willing to pull two shifts, then (hell) a third,
To save a session from a deadlocked state;
At times, (to put it mildly) unrestrained--
Almost, at times, a nerd.

I grow old...I grow old...
dBase II and Wordstar are no longer sold.

Shall I start a BBS? Do I dare to try to teach?
I shall take my palmheld portable and hack upon the beach.
I have heard the networks passing packets, each to each

They have no traffic for the likes of me.

I have seen the Altair live and die
And software startups score on sorry score--
And millions made by men like Mitch Kapor.

We hackers linger by our leading edge
Forgetting what is pending in the cache
Till practice hurtles past us, and we crash.

The Zero-G Polka

Originally published in *PyroTechnics*, an SF fanzine, late 1970s

(To: "The Beer Barrel Polka")

Think of action, and reaction
Never mind the lack of traction;
When the band begins to play,
Grab your partner's hand,
Simply kick away.
We'll be whirling, we'll be reeling
And rebounding off the ceiling--
Don't pull my leg, my dear;
I haven't one to spare!

When Stashu's saxaphone begins to blare,
You can tell he pumps a lot of air;
He tends to fly around the ballroom;
That's why we've tied him to his chair.
So if you'd like to join me in a beer,
We could meet above the chandelier,
And suck some Schlitz from plastic bottles
And sing: *Hurray, the gang's all here!*

Chorus:

Turn off the G-force,
We wanna dance in mid-air!

No other recourse;

Cling to the floor if you dare!

But let us warn you:

A foot in the face is no crime--

For--

They say all is fair in love, war, sex

And two…fourths…time!

I Should've Been a Jedi

From *Jeff Duntemann's Contrapositive Diary*, May 8, 2016

(To: Toby Keith's "I Should've Been a Cowboy")

I'll bet you never heard ol' Luke Skywalker say,
"Princess Leia, have you ever thought of runnin' away
And settlin' down, would ya marry me?
(or at least get me the hell away from Tatooine...)"

She'd've said "yukkh!" in a New York minute;
Incest's against the law; there's no future in it.
She just stole a kiss as they swung away;
Luke never let his hormones…get out of place.

Refrain:
 I should've been a Jedi
 I should've learned there is no "try…"
 Wavin' my light saber, knockin' the arms right off some ugly guy.
 Blowin' them Empire ships
 Right out of the sky;
 Nukin' those Death Star cores;
 Yeah, I shoulda been a Jedi.

I mighta had a sidekick with a fuzzy mane,
Flyin' blind by the Force, just like Ben explained.
Takin' potshots at a Tusken Raider;
Givin' a hand to your daddy Vader…

Blast off, young man, ain'tcha seen them flicks:
Outer space is full of rayguns, wookies, and chicks!

Sleepin out all night inside a tauntaun's guts,
Dreamin' 'bout the stars instead of freezin' my butt…

(Refrain)

In some ways, this edged close to the problem I had with "Wagger Central" (see above) in that it was a parody of a country-western song. A lot of people (including many I respect) simply cannot abide country music and certainly don't listen to it. Fair enough. I don't listen to rap. Maybe I shouldn't have worried; according to Billboard, it was the most-played country song of the 1990s. It's one of my all-time favorites. And the parody has gotten me some fan mail, even though it has only appeared in a blog post, until now.

PART 4. MEMOIR

Don't expect to see a published autobiography from me. I am not (and in truth, don't want to be) famous enough to warrant one. That said, I have written a great deal about my life over the past ten years, but it was for an entirely different reason: I'm closing in on 69 as I finish up this book, and I want to be able to remember as much of my life as possible when I'm 90. So I might as well write it all down now, while I and my sister (who has a better memory than I do anyway) are still alive and mostly functional. The previous generation are now (with only one exception) in the hands of God.

So why don't I just publish my memoirs anyway? Well, a lot of it are lists of things, like the addresses of places I've lived; cars, computers, and dogs that I've owned, and so on. And I hate to say it, but it's true: By conscious choice, I've lived a comfortable and mildly boring life without a great deal of drama. I had wonderful parents, I met my soulmate when I was barely seventeen, and opted out of the sorts of adventures that have gotten a lot of people in *way* more trouble than I ever wanted to deal with.

The worse problem is that a life is a tangled sort of thing, and teasing it all out into a linear narrative may be a fool's errand. In truth, not all stories I could tell stand on their own. Everything refers to everything. The cast of characters is large. And the world I grew up in is now decades in the past and incomprehensible to those who weren't even here on Earth until the twenty-first century.

In this section are a few snippets that stand well on their own. Some few I published on Contrapositive Diary. Some go back a long way (all the way back to my first professional sale as a writer) while others come out of the more recent memoirs project that I began in 2012, when I turned 60.

All the World's a Junkbox!

From 73 Magazine #171, December 1974

Originally Titled "How to Get Zillions of Parts for Nothing"

Ya gotta have a junkbox. I mean, there is nothing more embarrassing to a basement experimenter than beginning a new gizmo and discovering he hasn't a 47K, half-watt resistor to his name.

I found myself at just about that point not long ago. The American Radio Relay League recommends politely wheeling and dealing local TV repair shops out of burned-out chassis for a dollar or two apiece. Value for value, they say; after all, the poor guys are out to make a buck.

So I spent an afternoon visiting local repair shops with a buck or two in my pocket and an innocent look on my face.

One man politely told me he didn't run that sort of shop; another said old chassis attract cockroaches. One guy *did* offer me a 1957 RCA chassis minus tubes, tuner, CRT and half of everything else — for five dollars. I said no thanks, went home, and began composing a thoughtful rebuttal to the League article.

Little sister WN90VO wandered by. "No luck, huh?"

"Value for value," I kept mumbling. "This country oughta go on the junk chassis standard."

"Put an ad in the paper," she said. "Begging busted TVs from TV repairmen is like buying sand in the middle of the Sahara."

She had a point there.

The ad read: WANTED: BROKEN RADIOS, TVs, PHONOS, ANY JUNK ELECTRONICS NEEDED BY YOUNG RADIO AMATEUR FOR EXPERIMENTATION. I WILL HAUL AWAY. CALL XXX-XXXX EVENINGS.

I placed it in a local supermarket-and-drugstore ad flyer with a circulation of perhaps two thousand middle class families within ten blocks or so. It cost me fifty cents, and I expected half a dozen old TV sets and maybe a clock radio or two to cannibalize.

Once again, I had underestimated middle-class America. By suppertime of the day the flyer hit the mailboxes I had eleven calls scrawled on the back of a pizza board tacked to the wall by the telephone. Lots of junk in the neighborhood, apparently. I thought it was funny. The next morning Little Sister and I borrowed my father's station wagon and began the rounds.

We put it all in the garage. There was no other place to put it. Every respondent, it seemed, had a black and white console TV set gathering dust in the basement which he was too old/lazy/busy to cart out in front on garbage day. Most were nice old ladies who approved of my conservative haircut and wondered what on earth I was going to do with all that junk.

Never use the word "ham" in channel 2 land. I had a little speech about preparing myself for a useful career in electronics through construction of small transmitters and receivers. They liked that, and marveled that I wasn't out on the streets breaking windows like most college kids. And the calls just kept coming in.

They came in hot and heavy for almost a week. The garage was filling with alarming speed. The callers began to offer not one TV but two or three. One pleasant old gentlement gave me five, adding that he couldn't see too well anymore and anyway, there was some (expletive deleted) radio ham down the street who always messed up the picture. He told me he was glad I was going to be a disk jockey.

I began to lose track of some of the calls, forcing the callers to call back, asking if I had forgotten. One persistent woman called me five times until I emptied her basement of a TV and three grungy phonographs. The calls occasionally got a little weird. One lady with a raspy voice asked if I wanted to buy two manglers for 25 dollars. I figured a mangier was a 300 watt CB linear or something, but had the curiosity to ask. A mangle (bless her heart) is a 200 pound rotary ironing machine that literally squishes the wrinkles out of things. Producing a lot of TVI too, no doubt. I told her I was broke and hung up before she could offer them to me for nothing. Another chap had five hundred three-transistor radios to sell in a hurry for a hundred bucks. He refused to give me his phone number and is probably still at large.

Contributions were not always broken. Two of the TVs worked excellently, and I donated them to apartment-hunting friends who enjoy the mind-rot machine. One sour fellow handed me a 40 watt tube-type stereo amp, and told me it had worked fine for years, but recently had begun blowing the house fuse every time he plugged it in. He thought I might be able to get a few parts out of it. I looked down at the line cord and noticed that the insulation had crumbled right where the cord entered the cabinet. The wire had been twisted, and…of *course* I could get a few parts out of it; thank you, sir. A little soldering-gun work and it's been pumping John Denver into my speakers beautifully ever since.

Another man gave me several working tuners and amps which were "just cluttering up the house." The only cost to me was half an hour spent complimenting the bass response of his new system. Value for value? You bet!

Perhaps the best deal of all came from a retired gentleman who led me to a basement corner and pulled a dusty bedsheet off an enormous 1937 Zenith all-band floor-standing receiver, complete with magic eye tuning indicator and flawless dark wood cabinet.

"Bet you'll have some fun ripping this ol' bugger apart," he said to me with a grin. I agreed and carted it home. Just for kicks I plugged it in behind the garage, expecting it to blow itself to kindling. Instead, with the antenna lead clipped to an aluminum ladder, I copied a VE7 on 20 meter CW, using QRM for a BFO. No trace of AC hum. And a tremendous bass response, which is wasted on our gutless AM broadcasts.

A similar Zenith, needing only a filter capacitor, came to light about a week later. I have gotten fantastic offers for both of them from the antique radio freaks.

Nor were the giveaways limited to home entertainment devices. An elderly ham spent half an hour picking through his junkbox, filling eight boxes with 1625s, substitution manuals, ancient transmitting variables, relays and more than 200 pounds of power transformers, modulation transformers, and bathtub capacitors.

A second ham gave me an old but spunky Knight T-50 transmitter. A third sold me a mint-condition Central Electronics 10B exciter and 458 vfo for ten bucks, telling me to "get the heck off of CW." That was the only thing I paid a penny for.

It went on and on. I answered more than 50 calls, which continued drifting in for better than 5 weeks. Of those 50 I visited 36. The final box-score (kept with painstaking accuracy by WN90VO) turned up as follows: 31 broken TV sets, 2 working TV sets, 19 broken clock radios, table radios and transistor radios, 7 working clock radios, table radios and transistor radios, etc. 2 salvageable "antique" type radios, 3 unsalvageable "antique" type radios, 8 broken radios, 3 working amplifiers, 3 broken amplifiers.

Also, 4 working tuners, 1 broken tuner, 2 broken eight-track tape players, 4 broken intercom sets, 3 usable speaker cabinets with speakers, 1 working ham transmitter (not including the 10B), 1 broken photoflash strobe unit, 1 broken oscilloscope, 1 working 650V power supply, several old Spike Jones records, about a dozen boxes of loose parts from a ham and a man whose son had once played with "that stuff."

It took about eight weeks (i.e., most of the summer) to reduce all that junk to its component parts. I have a fairly respectable junkbox now, although I admit I have a few more 6AL5s and 5U4s than I'll probably find use for. But I saved all the deflection yokes, pried the copper out of them, and got 23 bucks for the lot. Beats hoarding pennies any day.

We're still crunching resistors out in the garage, and I suspect that the mice in the foundation have nests woven of greasy hook-up wire gorged by the pound from the bowels of yesteryear's boob tubes. My mother took a call on the ad as recently as Labor Day. She told the nice man I was out of town, and warned me that if I so much as *thought* about doing it again, I had better be out of town — if I valued my skin.

So, you poverty-stricken squawkbox-builders out there, I would recommend ignoring the League's suggestion to con TV repairmen out of totally blitzed TV chassis for "a dollar or two."

Why buy sand in the middle of the Sahara? All the world's a junkbox, OM!

Dig in!

This was the very first piece of writing I ever sold for money, in fall 1973. I don't recall how much editor Wayne Green paid me. $50? Sounds about right. If that doesn't sound like much, remember that this was almost fifty years ago, and that it was my Big Break. $50? No problem!

Just a few weeks later I sold my SF short story "Our Lady of the Endless Sky" to Harry Harrison for his *Nova 4* anthology. The book came out in summer 1974, so technically the story was my first published work. Paid more, too: $195. Wayne Green sat on "All the World's a Junkbox" until November of 1974, when the December issue of *73* hit the mailboxes. And the screwball changed the title to the execrable "How to Get Zillions of Parts for Nothing." Forgive me for reverting the title here.

Eating Irish

From Contrapositive Diary for March 17, 2013

As I passed the photo of my godmother, Kathleen Duntemann 1920-1999 on the bookcase earlier today, I quietly wished her a happy St. Patrick's Day. (It might be customary to say, "Wherever she is" except that I know *exactly* where she is.) She and my grandmother Sade Prendergast Duntemann were excellent cooks, and on St. Patrick's Day there would almost always be corned beef and cabbage, duck, or goose, all cooked using ancient Irish recipes. I'm not sure if it was a purely family eccentricity, but back when I was still living at home, a well-picked winter goose or duck carcass would be tied with some twine to a branch of the big sycamore tree by the back door. The birds feasted, and according to my mother, the fatty leftovers allowed the now-lean birds to survive the remainder of those nasty Chicago winters.

Of course, by May 1 there were half a dozen bird skeletons swinging in the breeze, which must have made the neighbors wonder.

As much as shamrocks entered into the spirit of Old St. Pattie's Day at our house, I never once heard my aunt or grandmother suggest that the famous Trinitarian clover was itself food. Finally, at age 60, I realize that they were–and apparently still are. Dermot Dobson posted a photo on Facebook, of a bag of shamrock-flavored potato chips. (Or crisps, in UK/Irish parlance.) Although I initially suspected that the crisp makers were being metaphorical and perhaps having chives stand in for shamrocks, when Dermot posted a shot of the ingredients list, begorrah! Those little green things are actually pieces of genuine Irish shamrock.

Of course, this is not an ancient recipe; Keogh's introduced the product only last year. A little research showed that the Irish evidently ate shamrock, though the implication was that they ate it in lean times when there wasn't much else on the menu. Shamrock is, after all, a species of clover. (We're still not entirely sure which species, of eight or nine contenders, that St. Patrick used to convert all those pagan chieftans.) I barely eat vegetables at all; I can't imagine eating what might as well be grass.

Whoops. Not only can I imagine it, I *remember* it: When we were eight or nine, the kids in my neighborhood would chew on what we called "sour clover," which was a local weed that could be found under most bushes. Many years later, while doing some yardwork for my mother, I found a sprig, chomped one of the three-lobed leaves, and felt that sharp sour tang. This time I looked it up, and found that our sour clover was *oxalis montana* (wood sorrel) which looks precisely like the quintessential Irish shamrock. It's sour because it contains toxic oxalic acid. Eat enough of that, and you will not only be eating a metaphor of the Trinity, you may in fact get to meet the Trinity face-to-face.

Ok, that would be a *lot* of oxalis, and as best I know we all survived the adventure, Irish kids, Polish kids, Italian kids, and mongrel kids like me whom God stuck together from a box of odd ethnic parts. To us St. Patrick's day meant that winter was almost over, that Bud's Hardware Store had just gotten in its first shipment of Hi-Flier kites, and that the color green would soon return to the Chicago spectrum.

When Aunt Kathleen died in 1999, an Old Catholic woman priest sang the Irish Blessing before her casket at the cemetery chapel, and I smiled to think of all those goose carcasses, and how Aunt Kathleen would as likely as not be hanging with The Big Guy Himself, driving snakes out of the neighborhood in her Pontiac, hoisting a glass of good Irish whisky, and keeping the kitchen warm for anyone who might stop by.

Live life in the active voice, this day and always. It's the Irish way.

The Lost Hobby of Microscopy

From Contrapositive Diary for March 23, 2013

Carol found some very small insects crawling around on Dash's neck yesterday while she was brushing him. She dropped several of them into a pill bottle followed by some alcohol. These were tiny bugs; I'm guessing the biggest one wasn't quite two millimeters long, and most were at best a millimeter. We squinted and used the magnifying glass that I keep in my desk drawer, and the best we could say is, *Yeah, that's a bug.*

I knew what I had to do next, and it took me *way* back. For Christmas when I was eight (the end of 1960), my father bought me a microscope. It was small and lacked a fine focus knob, but it had an iron frame and decent optics. For the next two years until I discovered electronics, looking at very small things was one of my main hobbies.

My father helped me get the hang of it. He had owned a simple microscope himself in the early 1930s, and I still have it somewhere: A black crinkle-finish tube about five inches high, with an eyepiece at the top, a slot for inserting slides, and a tilting mirror in a large milled cutout toward the bottom. He bought me a book called *Hunting with the Microscope*, by Gaylord Johnson and Maurice Bleifield (1956) and I spent a couple of years hunting for all the microscopic things the authors had painstakingly drawn on its pages.

Many of the drawn microorganisms were said to be found in rivers and ponds, and my friends and I haunted the banks of the Chicago and Des Plaines rivers in the summer with mayonnaise jars in hand, scooping up slimy water and the even slimier mud on the riverbottom beneath it. Holding up the jars against bright light showed them to be absolutely crawling with minuscule thingies in constant motion. I had a well slide and managed to corral some of the little monsters in it, but they didn't slow down long enough for me to identify them. None followed the corkscrew path that paramecia were said to exhibit. We saw no volvoxes nor stentors, cool as that would have been. Water bears too were AWOL. Most heart-breakingly, we never cornered an amoeba, which we longed to see eat something by engulfing it, which would be akin to watching *The Blob* in miniature–always a draw for ten-year-olds.

No, most of the critters that moved slowly enough to identify were micro-scopic worms. When my mother heard us talking about worms from the corner of the family room when my friends and I were gazing into my microscope, she made us dump the mayonnaise jars into the toilet and wash our hands. My mother was an RN, and although we didn't learn it first-hand until we were 13, both rivers were flood relief for Chicago's and suburban sewers. After even a modest rain, runoff would cascade from overflowing sewer mains right into the rivers, carrying raw sewage with it. So these weren't exactly earthworms we were watching. (Finding condoms tangled in the tree roots dipping into the Des Plaines river—before we knew what condoms were—was not a helpful clue.)

I'm honestly not sure whatever became of my little microscope. The good news is that Carol received a much better one she when was fourteen (a Tasco 951 with a fine focus knob) and earlier today, I pulled her microscope down off the high shelf and set it up on the kitchen island where the light was good. I looked at a few of the pickled-in-alcohol bugs, but they had been picked off Dash with a tweezers and were not in good shape. We cornered Dash and hunted until we spotted a live one. I carefully snipped the little tuft of hair to which the bug was clinging, and with some prodding managed to tack the bug to the sticky strip on a white Post-It. (Gaylord Johnson would have been proud.) Under the microscope, it was unmistakable: *Linognathus setosus*, the dog louse. The tacky Post-It strip kept it from walking around, and we were able to see how it clung to a strand of dog hair with its hooked legs.

Dash got a prompt treatment with the usual doggie bug meds, and in a day or two whatever lice remain will be gone. In the meantime, I have to wonder what happened to the microscopy hobby. Astronomy and electronics are both big business, but beyond some Web sites I don't see much to indicate that anybody is digging through river mud looking for water fleas anymore. The instruments are cheap compared to good electronic test equipment or telescopes. You can get used stereo microscopes on eBay for $250 or less, and used student microscopes like Carol's for under $50. Rivers are a whole lot cleaner than they were fifty years ago, and I'm thinking that if I sampled the Chicago River today I might score a stentor or two, and maybe even an amoeba. Granting that Google is a much better way to identify the stuff you're looking at, I might order a copy of *Hunting with the Microscope*, just for fun. No, I don't really need another hobby, but I want to be ready the next time something really small comes calling and I need to know what it is.

New Year's Eve, 1958

From *Good Times*, the Santa Cruz free paper, December 1988

I heard my old man come down the stairs from the upstairs bedroom. I was awake, and was shining clown-faces on my bedroom ceiling with a ridiculous toy flashlight that I had received for Christmas and unaccountably loved.

He cranked the doorknob and peered in. I expected quiet orders to "hit the rack, dammit!" but, remarkably, he grinned his slightly crooked grin and said, "Come on out and toast the New Year with me."

So I slid out of bed and skittered into the kitchen on bare six-year-old feet. I took my usual place on the broom-closet side of the kitchen table. The old man pulled down two crystal glasses with long stems from the high cabinet, and placed one on the battered Formica in front of me.

I wasn't sure what to make of it. He had tucked us into bed hours earlier, dressed in his at-home T-shirt and drab baggy pants. Now he was in his best blue suit, high starched collar, and dark red tie with the tiny slide-rule tie-tack. My mother would be working all night at the hospital, and my little sister was still fast asleep in her crib.

The kitchen was mostly dark, lit only by the bright colored lights circling the Christmas tree in the front room. It was dark enough to see the orange glow from the tubes inside the radio on the kitchen counter. Somebody was talking on the radio, not quite loud enough to understand over the refrigerator's wheezy clatter.

My old man yanked the refrigerator handle, and for a moment the single bulb within was blinding. He pulled a tall bottle from the rack on the door, turned, paused with the door half-closed, then yanked it back open and pulled a can of Nehi Grape from the top shelf. I watched him fiddle the foil and the wires from the tall bottle, and we laughed when the resounding *pop!* shot the cork across the room.

He pulled a church-key from the junk drawer and opened the grape soda for me. It was hard enough to score a Nehi during the day (and *never* during dinner!) and here he was pouring fizzy grape soda into that strange tall glass in the middle of the night.

That done, he filled his own glass from the tall bottle. For a long moment, we waited in silence. He had not touched his glass, and I left mine longingly alone, assuming that this was one of Those Occasions. Hank was sleeping with his mongrel rump plastered up against his favorite heating register, and everything seemed very warm and safe if only a *little* bit strange.

"Howcum you're all dressed up?" I asked. That was the mystery at the center of it, I was sure.

"You ever felt afraid of the future?" he asked.

I shook my head. The future, to me, was full of rocketships and space stations, and the Good Guys always blasted the aliens in the end, right? What was a little scary was my old man the engineer answering one question with another.

"Always look the future straight in the eye," he said, with a sudden distance that frightened even more than his answering question. "And wear your Sunday best, so it'll know you mean business."

In the silence that followed, I heard voices counting down on the radio. All at once, the wordless cheers told me it was New Years. Down the block the big kids were setting off firecrackers. Hank twitched a half-terrier ear and went back to sleep. Down in the basement our tired old furnace ground into roaring life.

And the old man was back from his distance, holding the glittering glass high in the air. "Happy New Year, Duntemann," he said with that paradoxical, loving drill-sergeant's voice that I will miss all the rest of my days. His old-style rimless crystal bifocals flashed in the Christmas lights, his ice-blue engineer's eyes again smiling that omnipotent smile. I cannot forget his face at that moment because it is *my* face, I who am now exactly as old as he was at the end of recession-year 1958.

"Happy New Year!" I said too loudly in reply, holding my glass in his direction in imitation of his gesture.

He raised his glass to drink, and I was already draining mine before I noticed that his never quite reached his lips.

Instead he had turned toward the empty corner of the kitchen and held his glass in a toast in a direction that was not toward Mother at the hospital, nor toward my sister in her crib, nor anywhere else, but instead in a direction that I never understood.

Until now.

Don't ask me to read this out loud. I can't do it. (I've tried.) I wrote it as my entry to a holiday essay contest put on by the Santa Cruz free paper, *Good Times.* It was the first free paper that I ever read on a regular basis, and turned out to be a far better written (and far less gonzo lefty) than I had been warned. It included a lot of excellent comics, especially Shari Flenniken's "Trots and Bonnie," but also the very weird running strips like Charles Burns' "Big Baby" and David Lynch's "The Angriest Dog in the World."

This incident really happened, pretty much as I relate it here. I admit that my memories of my six-year-old self have gotten pale and may be garbled in spots. I also admit that I had to re-create my weird conversation with my father, which may have had some irrelevant twists and turns.

I was definitely outside of my longstanding middle-American cultural context. That being the case, I was boggled that I won second prize, which was a dinner at a very nice local restaurant. Several people looked me up in the phone book (remember phone books?) and told me how much they liked it. One gentleman told me it reminded him of his father, who had died young—as mine had. I heard his voice start to break, and he said, "That's all I wanted to say. Thanks. Good bye."

My First Date, Relatively Speaking

Original essay from 1994, published here for the first time

My first date, relatively speaking, was Mary Kate McGuire. That is, Mary Kate was a relative (the oldest daughter of my father's youngest cousin) whom I took to a dance at the behest of other relatives (like her parents and mine) who thought it would be a Good Thing.

Little did they know.

Consider what I was at age 14: A geeky clotheshamper of a kid with disheveled hair, an overactive imagination, and little in the plus column except straight A's and skin with some strange genetic immunity to pimples. All through my teens I got one pimple a year, like clockwork, usually in some place where nobody could see it, like the inside of my right thigh. I guess that was God's little compensation for my being born completely lost at sea in terms of connection with my own peer culture.

It wasn't like I was a physically ugly kid. Actually, I was blessed with completely average looks that stayed average until my hair fell out, at which time I declared the whole question meaningless. No, the trouble was that I completely lacked that low-level telepathy that allows kids to somehow know without asking what's in and what's out and whether or not blue shirts go with brown pants. My engineer's solution was to request one color pants (dark blue) and one type of shirt (a weird permapress madras) that nominally obeyed the rules, and then wore them all the time. They were always on sale at Goldblatt's, which was a big plus.

Add to this a lack of rhythm, disinterest in sports, and a broad contrarian streak, and you have the sort of shyness that truthfully comes from having no idea what to say in any given situation. Everyone else in my peer group (except for my fellow nerds) seemed to be following some sort of script written in an alien language.

I was invited to none of the forbidden "mixed parties" the nuns railed against all through eighth grade, even those my best friend Art attended—Art, whom I always considered as nerdy as me. He might have had trouble following the script, but he could at least *read* it—and did cooler things like learning guitar while I stuck with piano. It was perplexing to hear those I thought I understood rhapsodizing about a chance to wear those new "tips, heels, and tapers"—pointy-toed Cuban-heeled boots with tight pants—while I remained loyal to my unchanging blunts, flats, and baggies.

In fairness to my classmates, there was a time or two when they reached out to me—like the time Cathy Ceglarek brought me a chicken leg at the eighth grade class picnic, when I was sitting off by myself in a blue funk. That incident threw me into a complete double panic: I had had no idea it was coming and then no idea what to do about it. Whatever fondness she may have had for me was lost in the total muddle of how I tried and failed to respond.

And that was the core of the problem: I was without any model for interacting with my female peers in an acceptable fashion. I wasn't rude or habitually sullen, but energetic, randomly light-hearted and probably considered insane by the genetically hip.

Fast forward to May of 1967. I was a high school frosh in Army Junior ROTC, where the Powers had decreed that there would be a Military Ball, and that All Of Us, frosh pointedly included, would be required to attend.

A military ball was a sort of prom with brass buttons. We had our uniforms, courtesy the Army, so renting a tux was a thankful nonissue. But the other required ingredient—a girl—was not so easy to obtain for certain wild-eyed 14-year-olds with hair sticking out in all directions at once.

The old man told me to stop worrying, joking that if the Army wanted me to have a girl, they would issue me one. He was right: The old man, always the soldier and at whose urging I had joined ROTC to begin with, huddled with his relatives and came up with the solution: Mary Kate McGuire would be my escort.

Mary Kate was a year younger than I, an eighth grader living out in the burbs on the periphery of O'Hare Field. I knew her vaguely but not well enough to call her a friend; we spoke kind of lamely when the families came together. She was pretty without being frighteningly so, and had that scrubbed and wholesome look that could have come off a Norman Rockwell *Saturday Evening Post* cover, complete with a scattering of freckles.

Going on a date with one of your relatives seemed at some level ridiculous to me, all the more so by the fuss my mother was making over the whole business. She spent a Saturday afternoon teaching me how to dance. Two-step, well, OK. Box waltz, egad—can I draw a diagram? And the jitterbug…Mother, please, they don't do things like that anymore! She had bought umpteen rolls of Super 8 movie film to record the occasion. I got the electric shaver I was going to get for my fifteenth birthday a month early so I would look my sharpest—even though for me, five o'clock shadow still didn't show up until Tuesday.

Mother persisted in ordering flowers from a local florist with a daughter my age who knew who I was and doubtless spent plenty of time snickering with her girlfriends over the very idea that Jeff Duntemann would be going to a formal with a *girl*...

Hey, when you're a nerd, dying a thousand deaths is something you do every day before breakfast. By that time I was pretty good at it. I also knew that if the old man decided something was good for you, dammit, it would be done.

Then came the fateful Saturday afternoon, when my dad drove me out to Des Plaines to pick up my "date." I rang the back doorbell on the wonderful 75-year-old house the McGuires lived in, still grumbling a little inside at the incongruity of it all. Mary Kate's mother answered with her characteristic cheeriness, and I stood there in the kitchen until Mary Kate came down the hall and into view.

It was a tectonic discontinuity that probably hit 7.65 on the Richter Scale. She was, well, *beautiful*, in a knee-length pastel-blue dress and white pumps. Chalk it up to not seeing her for a year or two—at 13, things happen quickly. But more disconcerting than that, I think, was the strange notion that she had done herself up so nicely for *me*. I mean, me, the mad scientist of Immaculate Conception grade school. How could this be?

But it was. And downtown we went, to the Grand Ballroom at the Conrad Hilton on South Michigan Avenue. I introduced her in the receiving line (without mentioning minor details like our degree of consanguinity) to the retired Army sergeants who taught ROTC at Lane Tech. I introduced her to the seniors, who as officers of the school brigade had the potential to make life miserable for the frosh who didn't do that required bit of hierarchical sucking up.

And I introduced her to my fellow frosh, many of whom—good grief!—had come to the ball *by themselves*, with no girl at all. With some uncataloged astonishment I realized that I was by no means at the bottom of the heap of grubby frosh nerdhood. On my arm was a real live girl, where before had been nothing but empty air and wistful fantasy to fill the loneliness that puberty leaves in its untidy wake.

Once the formal dinner was done and the tables cleared away, the music be-gan. Far from being the stodgy old grown-up stuff my mother had assumed they would play, it was a rock-and-roll band that played the same stuff we heard on the radio. With relieved (if clumsy) abandon, I danced with Mary Kate what my Irish grandmother testily dismissed as "wiggle-arse dancing." I was quite certain I was no good at it whatsoever—but it was just as clear that if Mary Kate knew, she gave no sign. And she was a goddess, with poise and perfect rhythm and a luminous smile. The discontinuity continued as I spied my dateless frosh friends sitting around the tables, looking enviously…at *me!*

Then the tempo went down hard, and the band began the emblematic "slow dance" song of the spring of 1967: "My Girl." There was a trace of irony (and a titanic slug of courage) in my asking Mary Kate to dance it with me. Without a word she took my hand, and my other hand went around her back. The respectful distance my mother had said all "good girls" required vanished as she pulled her-self close to me, and put her head on my shoulder. I was thinking desperately that I should be counting beats, or watching where my feet were going, or paying atten-tion to something other than this miracle of miracles: There was a girl in my arms, a real live girl and a beautiful one at that, and she was there willingly, and although she could not be and would never be "My Girl," she swayed to the music with me while an astonished realization rose from the deepest core of my adolescent soul:

I can do this!

The words rang in my mind, golden and luminous, while the music ended. As surely as Mary Kate would never be in my arms again, I knew that someday another girl would, and Mary Kate and I chattered happily in the back seat of my father's car on the long drive back home out the Kennedy Expressway. As best I could tell she had had a good time, and I was filled with the intoxicating feeling of having gotten a handle on something that had always eluded me. I now knew it could be done. All the coaching my dad had given me: "Be polite, be kind, ask her how you can help her, try to be her friend" had always bounced off of me in a despairing funk. How could I ever say such things to a girl? I didn't know the first thing about girls.

Now I did.

No, I didn't kiss Mary Kate goodnight that night. She wasn't mine to kiss, and that had never been the point anyway. It took me twenty years to figure out that all the fuss my parents went to over that silly Military Ball was meant to provide their only son with a model—a template—for taking his first halting steps into a world of adult rituals, work, and responsibility. And later that year, when a girl named Judy asked me if I would kiss her, my knees shook but her pale young lips met mine as smoothly as lips in first kisses ever do. What might have been inconceivable had become simply the next step in a process that I finally realized—with unwitting help from the daughter of my father's youngest cousin—I had every right to partake in.

I saw Mary Kate only two or three times after that, and while she smiled and talked politely to me, the magic had done its work and passed on, and we were both headed in our own directions toward adulthood. I wonder sometimes what she told her girlfriends about the Ball that following Monday, and the weird guy her mother had set her up with, second or third cousin or something, with hair that wouldn't lay down for nothing. I'm sure they asked her if I was cute, and I hope that in reply she told them I was nice, for that was her way, and I had tried my best to be. And when on that fateful day in 1969 I found myself seated next to a girl who (in my 17-year-old view) surpassed all horizons of human beauty, I swallowed hard and did what my father had taught me: *Be polite, be kind, ask her how you can help her, try your best to be her friend.*

It might be exaggerating to say that, seven years further down the road, Carol Ostruska married me because my second cousin Mary Kate McGuire showed me that it was possible. Yes, it might be…

…but only a little.

I wrote this in 1994, hoping to sell it to one magazine or another, but never even tried. On one hand, magazines that used to publish little memoir pieces like this had gone away even as long ago as 1994. On the other hand, I think it cut a little too close to the heart to show it to the world at that time. I suspect my subconscious was telling me to just tell the stories out of my life, and not worry (yet) about who, if anyone, would ever read them. This story thus became the seed that precipitated my memoirs, which now comprise over 100,000 words. I'm writing them not to publish but simply to record: I want to be able to remember the details when my memory for details starts to go.

It seemed unlikely that I would ever see Mary Kate again. She had lived in Iowa for quite a few years. When I was in Chicago I sometimes visited her mother, Mary Ellen McGuire, who was my father's youngest cousin. Then one day in 1997 when I went over to Mary Ellen's elegant 100-year-old house, someone else answered the door.

"Hi Jeff!" Mary Kate said, laughing. "Want a highball?"

We weren't kids anymore. I was 45 and she was 44. But we laughed and talked and hugged and became friends again. I don't think she completely understood how important the military ball had been to me. So I sent her this essay. Now she understands.

A Kingly Gift

(Excerpted from *Kick Ass—Just Don't Miss: The Life of a Contrarian Optimist*

By 1962, my grandmother Sade Duntemann was in ill health, losing weight, and a shadow of the wry, exuberant and very Irish woman who had sung mock opera from the railing of our staircase and taught me my ABCs in the mid-1950s. Sade had done something else: She encouraged me to tell her stories about the adventures of my stuffed animals, starring a nondescript little dog I loved greatly and called Baby Baah. (Baah still exists, though he was much patched during the 1950s, and he now has a place of honor on my library wall.) And so my four-year-old imagination learned that stories came from other places than books, including (what a notion!) *the back of my own head.*

Once I started, I never stopped. When I was 8 (1960) I wrote an adventure in longhand (again about my toy dogs) on yellow foolscap paper and presented it to her. She told me again and again how much she enjoyed it, and showed it to her friends and her Prendergast cousins on the South Side. Not quite two years later, she told me that she wanted me to have her typewriter.

It was an Underwood Standard #5, dating back to the first years she was married, probably 1921 or so. (It was thus older than my father.) It was the sort of typewriter people picture from old movies, especially those depicting Depression-era newsrooms. It was a huge wad of iron, so heavy I couldn't lift it myself at 10. It was my poor mother who lugged it out of Sade's basement and put it in the back seat of our '57 Pontiac Chieftan. I'm sure she was wondering while she puffed her way up the stairs to my room: *What in God's name is he going to do with this?*

She didn't have to wonder for long.

Sade herself hadn't done a great deal of writing on it, so it was in very good shape, having spent forty years in the basement of her house, mostly untouched and completely uncovered. (I cleaned up the inevitable dust with a rag and some Q-tips.) It took me a little while to get the hang of operating it, but I was good at trial-and-error and was off at a trot, typing first with two fingers, then four, and finally seven. Why seven? Simple: *It was enough.* By the time I was in sixth grade I was typing as quickly as the machine could handle, and I only needed seven fingers to do it. Since then I have managed to engage all but my two little fingers in the typing task. It's not anything like proper technique, and watching myself type just now I realize that probably 80% of the work is done by six fingers alone.

No matter. Typing was wonderful, nay, *magical.* With only a little work I could create pages full of stories that looked a great deal like the pages in the library books I was reading about that time, and the (fateful) Tom Swift, Jr. books that I bought for a dollar at Woolworth's because the local public library would not shelve them. Words written in my sloppy Catholic-school longhand were somehow inescapably *my* words…but words that came out of my typewriter looked like they could be anyone's words. Until I began typing my tales, I thought of them as written for myself—and perhaps for Grandma Sade—as ways to pin down the expansive fantasies that echoed around the back of my slightly lumpy little skull. Once I began typing them, I began to share them with my friends from Scouts, and later with Lee Anne down the street. Somewhere along the way, I began to dream that someday there would be something like a Tom Swift book, and that it would have my name on it, rather than the unlikely (and in truth, fictional) Victor Appleton, Jr.

It was a change that I couldn't appreciate at the time: Until 1962 I was a goofy little kid who told stories. After 1962, I was a *writer.*

Sade died in November, 1965. I missed her much more than I missed my other grandparents (two of whom didn't even speak English) even though she had her prides and her prejudices and had not always been kind to my mother. I forgive easily (what's the point of resentment, anyway?) and there are times when I shake my head while pondering how my life might have been different had she not believed in me when and how she did.

Firejammer

(Excerpted from *Kick Ass—Just Don't Miss: The Life of a Contrarian Optimist*

As I mentioned a little earlier, I had had some success with SF shorts in the magazines, but it was a narrow sort of success, and limited to precisely one magazine. I had sold two stories to George Scithers before we even left Chicago (if just barely) and a third a year or so later, in 1980. About that time, Joel Davis, the publisher of *Isaac Asimov's Science Fiction Magazine* decided to create a new title in the memory of the pulps of the 1940s and 1950s: *Asimov's SF Adventure Magazine*. George Scithers asked me if I would have anything to send him for the new title, anything with action and ideas and, well, more action and ideas.

Well, geepers! If I can't do that, well, what *can* I do? I told him I had a concept for a novella that would be perfect, and I got to work.

The concept goes back to high school, though as with my concept *Oxidation* I had done a great deal of conceptualizing but (by 1980) no writing at all. In high school I was a huge fan of Keith Laumer's novels, especially his light-hearted Retief adventures including *Retief's War* and *Retief and the Warlords*. Laumer's adventures always had goofy aliens, snotty repartee, non-stop action and boatloads of adrenaline. Damn, I wanted to do that!

The story was called *Firejammer*. The core gimmick was a planet in a complex system with lots of Earth-sized-or-larger planets. Tidal forces on the planet's crust kept the planet in a constant state of volcanic turmoil. In particular, there was a large valley between mountain ranges, with a fault-line crack running down the middle of it. Every 200-odd years, an alignment of planets caused the crack to open up, filling the valley with molten rock. The valley's inhabitants built castles out of stone—and when the Day of Fire arrived, the castles floated on the molten rock.

The castles didn't just float. They had sails made of chain mail, and in the furious winds raking across the liquid magma, they sailed and tacked and made war on one another.

The alien inhabitants of the valley were very Laumer-esque. They were big, powerful creatures with four legs, and had a vague resemblance to svelte-er hippopotami. Their front legs had hands, and they could walk as necessary on their hind legs or on all fours. Their huge jaws were strong enough to crush rocks. They had long forked tongues—and on the end of each half of their tongues they had four-fingered hands with opposable thumbs.

They were omnivores in a broader sense than humans, in that they ate meat, vegetables…and rocks. (The humans call them "rockchompers.") The rocks were not for nourishment but to provide raw materials for several bodily functions that we humans do not have. Most usefully, they excrete a sort of epoxy glue, which hardens when touched by their saliva. So, yes, the picture in your mind is accurate: When they need to glue something together, they shit on the ground and then spit on their shit. Stir well, and you can stick a civilization together with it.

The story begins when representatives from the Tripartisan Economic Combine (a very Trekkish galactic federation that had evolved out of my high-school future history of Intergalaxy) arrive to negotiate for a deal on their, er, glue. The Combine's point man, sent a year earlier to learn their language and culture, knows about the Day of Fire but doesn't alert his colleagues…because he has in mind an outrageous experiment: Attach a Combine orbital shuttle to the stern of one of the castles as a rocket engine, and see how the battle goes.

The story was not my first attempt at humor, but it was the first time I attempted humor through cultural references. The names of the several spacecraft are names from Watergate, (Nixon, J. Edgar Hoover, John Mitchell) which in 1980 was still very much on the public mind. One running gag is Icehall providing Turkey with a gimmick ebook: A dictionary of American slang and colloquialism from 1620-2000. I put a lot of 60s and 70s catch-phrases in Turkey's mouth, like "It's been years since I fit into a ten," "Crazy world, ain't it?" and so on. Turkey thought it was the way Americans talk all the time, and basically memorized the whole book. When speaking his own native tongue (which I put in parentheses, and included archaisms like thee and thou) he speaks like the highly educated heir-to-the-throne that he is. When speaking his dialect of American to Icehall, he sounded like a 60s hippie or a 70s hipster quoting TV commercials.

It was funny. My friends thought it was hilarious. But putting era-specific slang into a story is a mistake, as I learned later on.

When I mailed it in, George Scithers didn't precisely reject it. He said he enjoyed reading it, but I had taken so long to write it that by the time I submitted it, Scithers and Davis had decided to cancel the magazine for being a cash sink.

(Having later worked at magazines and run my own for ten years, I now know exactly what the problems were.)

Anyway. The story wasn't terrible when I originally wrote it in 1980, and it improved with time as I poked at it down the years. I pulled it out now and then, hoping that there would someday be someplace I could sell a 27,000-word light-hearted space adventure in the style of Keith Laumer.

Nothing ever turned up in traditional publishing. By the time we moved to Colorado in 2003, I had begun tinkering with self-publishing print-on-demand (POD) books through services like Lulu. I had high hopes for purely digital eooks, and watched early ebook readers like the Sony Reader, which I bought, and the Rocket Ebook, which I didn't. In 2007, Amazon introduced the Kindle reader, which basically normalized the idea of ebooks and put them squarely in the mainstream. I thought the first Kindle was a little clunky and didn't buy it, but I was soon reading ebooks regularly on my IBM Thinkpad X41, which was "convertible." This meant that the keyboard could fold around behind the laptop's body, creating a (somewhat heavy and thick) tablet. Reading on the X41reminded me that I had a 27,000 word short novel sitting on disk somewhere doing nothing...and ebooks don't need paper. With ebooks in mind I rewrote *Firejammer* yet again in early 2015, but was so distracted by moving back to Arizona in that timeframe that I literally forgot that I had rewritten it.

So things sat until 2019. We were finally done customizing our new house in Phoenix, and I pulled out Firejammer one more time. With a sigh I opened it to begin a strong rewrite to pull out all the little 1980 jokes that contemporary readers would not get. Yikes! I realized, boggling, that I had already done the rewrite four years before. *Firejammer* was (finally!) ready to be published.

It took me one long afternoon to lay it out in the Jutoh ebook editor. I commissioned a color cover for it from freelance artist Augusta Scarlett, and when the cover came in precisely as I wanted it, I ended a nearly 40-year odyssey. I posted the ebook to the Kindle store, and a (thin) paperback to Kindle Direct Publishing, the successor to Amazon's CreateSpace.

It has sold reasonably well. It's been reviewed favorably in several places. And it currently holds the record for time spent sitting in my trunk—which makes me wonder what else in the line of salvageable fiction may be wasting away down there. Excuse me while I go look.

The Nose of an Old Dog's Soul

(Excerpted from *Kick Ass—Just Don't Miss: The Life of a Contrarian Optimist*

Here it comes again: You're dreaming, and there you are, wandering around a phantasmagorical realization of your high school or college, desperately looking for English class. The panic is rising in the back of your mind as you realize you haven't been there for a great many years, have probably missed too many class days and tests to make up, and you will therefore never graduate, never get your degree, and ultimately spend the rest of your life living under a bridge or flipping burgers at Wendy's.

My conversations with any number of people indicate that what I've just described is, in fact, the commonest recurring dream in America, commoner than the oft-cited nocturnal bugaboo of abruptly appearing nude in public.

Recurring dreams fascinate me, because they are a window on what we truly fear. That nasty dream of returning to college or high school after ten, fifteen, or thirty years indicates more than just a fear of failure. It speaks of the terror many of us have of being unable to "work the system" of life, despite our best efforts and intentions. Interestingly, the only people who seem to have these dreams are those who have successfully finished high school or college and gone on to achieve a reasonable degree of success. Even in the face of triumph and against the certainty of a good many years, anxiety seems to well eternal. Almost all recurring dreams leave us with that lesson.

Almost.

By all counts the strangest recurring dream of which I've ever heard is one of my own. It departs powerfully from the universal anxiety theme of searching unsuccessfully for calculus class, or finding yourself in the wrong church on the morning of your wedding. Briefly: In this dream, I take an abrupt aside from whatever unlikely dream situation I may be in to wonder how old my old dog Hank must be now, and marvel at how old he's gotten to be. Why, he must be the oldest dog in the world by now...

To explore this peculiar night-time phenomenon, I'll first have to lay down some history. When I was very young my parents had a little blond cocker spaniel named Rebel, whom they found as a stray while my father was in engineering school in Georgia in 1950. Rebel died unexpectedly of lymphoma in 1956, when I was four. I don't remember much about him except his warmth and his furriness, and his gentle willingness to curl up on the floor with me and be grabbed and hugged with less gentleness than the poor creature deserved.

I do remember my heartbreak at his passing, and remember standing on tiptoe looking down into a chicken-wire pen at a farm-and-garden store somewhere in Des Plaines, where ten mongrel puppies romped and played under a sign reading, "Puppies $10." My father suspended his usual sound judgment by allowing me to pick the one I wanted. Perhaps I was afraid the active ones would be too lively for me. Perhaps I was showing some early sympathy for the underdog, as it were. But I chose the one sitting quietly in the corner, looking lonely and in need of love.

Love wasn't all he needed, as it happened. The poor thing had mange, worms, and numerous other parasites, and by the time my father had laid down his ten-spot, my puppy could barely stand. My mother was furious that my father had brought home an obviously diseased animal, but my father stood by and lived with his decisions. After considerable time and money spent in the care of Dr. Rudawski (and endless four-year-old bedside pleas to God of "Please don't let my puppy die!") Yankee came around and started to put on weight. I had a hard time getting my young mouth around the name "Yankee" for some reason, and called him "Hankie." Over the years, he was "Yankee" when my mother was displeased with him (which was often) and "Hank" nearly all the rest of the time.

Hank was nearly everything Rebel was not. He was independent, absolutely fearless, somewhat aloof, and extremely intelligent—easily the most intelligent dog I have ever spent any time with. Hybrid vigor had obviously deposited the very best that his beagle father and fox terrier mother had between them into a single package. He loved to roam the neighborhood, making sure in that time-honored doggie way that all the other neighborhood canines knew he had been there. My mother tried to keep him in the house and yard, but he learned early how to trip the iron latch on the front screen door, and always checked the back gate as soon as we let him out into the yard, to be sure some obliging meter reader or neighborhood child hadn't left the gate open. We piled firewood against the back fence until my mother spotted Hank tripping up the pile of logs and out over the fence into the alley; after that we piled wood up against the garage instead.

Hank was not a cuddly dog, as Rebel had been. He didn't play much, wouldn't chase balls, and would brook no gross physical indignities from me as Rebel had. In short, he wasn't a great deal of fun, and while we valued him for his abilities as a watchdog (he missed *nothing* that went on around him) my mother often wondered out loud whether his benefits outweighed his costs.

By 1966, Hank was ten and obviously slowing down. That summer my Aunt Josephine received two mongrel puppies from a neighbor. The puppies' mother was a miniature poodle. Who or what their father had been was unclear, except that he was a superb jumper and thought nothing of fences. My mother had always wanted a poodle, and consented to adopt the male of the pair.

Uncle Stanley had already named the puppies Smokey and Flame. They didn't look much like poodles, my mother protested, but I worked my magic again and persuaded her that Smokey was far too cute not to take home from Blue Island.

As Smokey grew, it became clear that his father had been big, black, and probably (Dr. Rudawsky said) a Labrador retriever. As the Smoker swelled to his full adult weight of 55 pounds, my father often joked to my mother's displeasure that his parents must have had an interesting courtship.

Smoke was another 180-degree turn in dog personality. He was big, dumb, lovable, playful, and infinitely tolerant of teenage horseplay. In the aftermath of Chicago's Great Blizzard of '67, our favorite game in the backyard was to have Smoke chomp down on my wooly stocking cap while I whirled around in circles in the middle of our snowbound yard, until he eventually tucked up his legs, closed his eyes, and spun in a circle until I let him go to tumble ass-over-teakettle into a snowdrift. He would instantly return, hand me my cap, and beg to do it again.

Smokey and Hank got along famously; when Smoke got too rambunctious, Hank growled his usual warning, and our junior canine immediately got the message. Hank, furthermore, took his role as adoptive father very seriously. In the summer of 1967, Hank was partly lame in the back legs and suffering from some obscure canine heart disease. I recall being at our summer home on Third Lake, out in the big front yard, with both Hank and Smoke sniffing around the flowerbeds in their usual way. About then, a huge blue tick hound came down the road (we were never quite sure who owned him) and crossed the yard to meet Smoker nose to nose. It was unclear what transpired between them, but the hound growled and snapped at Smoke, who backed away in his usual submissive fashion.

I heard another growl, and turned around. From the middle of the yard Hank was charging, lips pulled back, potent (if yellow) teeth in full display. His back legs weren't firing quite correctly, but he had that look in his eye indicating that he would defend his adopted puppy to the death if necessary. The closer he got to the blue tick hound, the lower the hound's tail fell, until with Hank's teeth barely a foot from his throat and closing, the hound spun around and bolted. Hank kept up the chase to the property line, though by then the hound was almost out of sight down Linden Lane.

Later that fall, Hank's hind legs finally refused to work at all, and he took that one last trip with my father to Dr. Rudawski's. I remember crying for awhile (which when you're fifteen is a difficult thing to admit) but Smokey was now my main dog, and while Hank would have sunk his formidable teeth into anything that tried to attack me, he had never been in any true sense a companion. Smokey, on the other hand, remained my enthusiastic shadow until I moved away from home at age 23.

I mean no disrespect to Hank here when I say that Smoke was my favorite dog from childhood, and remained the dog I recalled with most affection until I acquired the well-known Mr. Byte when I was 28.

I bore you with all this dog trivia because of the peculiar fact that I do not recall ever dreaming about Smokey. Not once, neither recently as an adult nor even when I was a teenager and he was curled up in the corner of my room. And for all the force of my legendary affection for Mr. Byte, (who slept at the foot of our bed every night for fourteen and a half years) I recall dreaming about him a mere handful of times, and only once or twice since his departure.

So consider this: At least six or seven times over the past ten or twelve years (and those are only the occasions that I remember) and as recently as the summer of 1998, I have dreamt about thinking about Hank. Note that I didn't dream *about* him, exactly. That is, he never shows up visually in a dream, wagging his tail or happily lifting his leg on something.

No. The recurring dream follows a very distinct pattern: I'm in some dream, doing whatever one does in dreams, which for me is usually wandering around alone looking at remarkable landscapes and monumental architecture. Suddenly I will turn away from whatever is going on, and think, "Gee, Hank's sure getting old! Why, he must be…he must be…" (and here I stumble over the legendary difficulty of doing arithmetic in dreams as I try to work it out from the years involved) "…well, twenty years old! That's real old for a dog." More recently, I got clever in my sleep, and thought, "Let's see, he was born the year my sister was, and my sister is 40 now, so he must be 40 too! He must be the oldest dog on Earth! Wow!" That done, the notion passes, and the dream resumes as though someone hit the "Play" button on a VCR.

Notice something truly remarkable here: I never for the slimmest moment entertain the possibility that Hank is dead. I always marvel at how amazingly old he's getting. That he is still alive is never in question.

This can't be due to an inability to accept his death at a subconscious level, as some of my friends have suggested. Shortly after Mr. Byte died, I dreamt of him walking out of the bedroom closet, and in the dream thought, "Well, the Byter is dead, so this must be a dream"—one of the very few times I have realized in a dream that I was in fact dreaming.

Let me summarize the distinguishing characteristics of this peculiar recurring dream of mine:

•   It interrupts another dream that is already in progress. This interruption is abrupt and in no way proceeds from the action of that dream; that is, I'm never seeing a dog or talking about dogs and move from there to thinking about Hank.

•   I never see him in a dream. I don't even dream about remembering how he looked, which is a slightly weird thing to consider but something I have done in dreams on occasion.

•   I never feel sorry for him, nor think about how he suffered his last few months, nor move off into other reveries about his life and adventures. My sole realization about him is that he's still alive and that he's been alive a long time for a dog.

•   I never consider the possibility that he might be dead. Never even once. In fact, the conviction that he is alive is surprising in its strength, from the standpoint of being awake and remembering my dreaming state of mind.

•   When it's done, the dream that I had been dreaming resumes without any change in direction. I don't move off to other things that involve dogs.

You'd have to be pretty dense not to see where I'm taking this by now. I believe strongly in an afterlife, and always have. The precise nature of that afterlife remains a puzzle (and something the Church tells us we shouldn't fret overmuch about it) but we have assurance from Scripture and sacred tradition that we will be reunited with all those who have died in the peace of the Lord. Well, will there be dogs there too? (Truly, it would be a dull kind of a heaven without them!)

Alas, the Church says almost nothing about the eternal fate of animals, and merely exhorts us to refrain from cruelty here on Earth. In my mother's severe Polish Catholic culture, it was taught that animals have mortal souls. We must take care of them and not cause them undue pain here in consideration of that mortal soul, but when an animal dies its soul dies with it. There was no talk of "dog heaven" for Rebel and Hank and Smoke.

More interestingly, the Theosophical tradition holds that there is a sort of collective soul for animals, one such collective soul for each species. When a puppy is born, its soul is drawn from the stuff of the Collective Soul of All Dogs, and when it dies, its soul returns to the collective, enriching it with its experience and the love it was shown while on Earth. So over time, we get better and better dogs.

Granted, this isn't a Catholic tradition—but something about it rings true to me. If I were God, I think I'd do it that way.

Now, what if, on death, a dog refused to return to that collective soul? That would be Hank's way. He did what we asked—if it aligned with what he wanted at the moment. When my dad called, he came, if only because my dad always had a Milk Bone in his pocket. Hank surely had his own agendas, whatever they might be, and he followed them with rare determination.

Recently I have begun to wonder if my recurring dream is the nose of an old dog's soul, briefly touching my mind while I sleep. Dogs don't have the power of language, and I doubt they know what they look like. So Hank wouldn't speak to me in words, nor show himself to me. About all he could do is plant a conviction in my head, with all the force of his indomitable will, that *I am alive*. And for a moment in my dreams, I turn aside and marvel at how old Hank is getting now. Why, he must be the oldest dog in the world…

Is he just being contrary, refusing to follow his Creator's will and return to the collective soul from which he came? Perhaps. Or maybe he has other things to do. My mother (whom I resemble in so many ways) dreamt frequently of demonic forces, and experiences I have had lead me to believe that such forces are real. I, on the other hand, do not dream of devils or demons. My nightmares remain on the level of searching for English class in my underwear.

I wonder sometimes if there is a brown-and-white spirit curled up nearby while I sleep, unseen but ever alert for evil things. Perhaps, from time to time, his upper lip pulls back to bare those formidable teeth and let that warning growl betray his presence. Were I a demonic force, I would find somewhere else to be in a *big* hurry.

Hank was certainly smart enough to have a canine sense of justice and obligation. In whatever way God allows dogs to be aware, he was doubtless aware that had I not chosen him from his chicken-wire pen that day, he would not have lasted out the week—and instead, he had twelve good years.

So perhaps, when my own appointed moment arrives, I will see him one last time. He will sniff my ankle to see where I've been, and wag his tail in approval. And I know what he'll be thinking:

Thanks, kid. You let me have a life, and I always take care of my own. You're here now. My job is done.

And with that, he will at last leap joyously into the Collective Soul of All Dogs, to contribute his fearlessness and loyalty to the raw stuff from which the Father will craft all dogs yet to come.

Dogs are important to me. Always have been. I've found it interesting that of the many dogs I've lived with, each had a distinct personality, with quirks unique to himself. The piece above dates back to 2000, and like the story of my date with Mary Kate, I had written it intending to submit it for publication to magazines that still publish such things, only to find that few such magazines remained. That was 2000; printed magazines are now so scarce as to seem an oddity, as was, in truth, the article itself. So a book called Odd Lots is the perfect place for it.

I consider this essay important because it was one of the seeds of the novel I finished and published in 2020: *Dreamhealer*. The novel is about repeating dreams, nightmares, the collective unconscious, and a guy who invades other people's nightmares and ends them, by banishing the little monsters who create nightmares for dreamers, and then feast on the terror those dreams evoke. Dogs do play a role in the plot: The little monsters (which I call archons) can be killed (and eaten) by dreaming dogs, of which there are many. *Dreamhealer* is what I call "suburban fantasy," in that it's set in the current day, in the sorts of suburban environments much despised by the literary set. Literary it's not. Fun, yeah. It's on Kindle, if you ever have some time to kill.

Part 6.
NONE OF THE ABOVE

Certain things defy categorization. Even in a book called *Odd Lots*, there are outliers. Here, to wrap things up, is a short collection of things that are so odd as to fit nowhere else. Everything here would probably be a blog post if I wrote them today. Or maybe not.

What all of the items in this category share is my insufferable optimism. As a college colleague told me decades ago: "The trouble with you, Jeff, is that you're too *damned* happy." Guilty. As for where it came from, I simply don't know. It's not like I went to Optimist Night School. It's just how I think. Yes, I've been lucky. Yes, I've avoided doing things that a one-eyed squirrel could tell would be bad for me. I belong to no tribes, having experienced the evil that tribalism engenders and knew well enough to throw it off before it got its hooks into me. I examine the hell out of my life to make sure that none of it was defined by anybody other than me. (This is another of the "gifts" of tribalism: Forcing opinions into people's heads and then fooling them into thinking that those opinions are their own.)

I reserve the right to say things like Good Wins. (I also knock out one of the o's and say it again, which is pretty much the same thing.) I believe that redemption trumps damnation. I believe in blind hope, and in James 2:24.

All of which may just be to say: I'm an optimist. Optimists are weird. Yes, they are—and I wear it now and will always wear it as a badge of honor.

Screwing Around on the Edge of Darkness

Sometime in late 1969, I wrote a pompous short called "Dialog on the Edge of Darkness." Unlike most of my high school short fiction, I still have it, and it's a groaner. In the story, Everyman (whom I called "The Wanderer") and Death walk along a dreary beach, arguing about why Death should not have The Wanderer's hide. The Wanderer loses every verbal ploy and seems doomed, until it occurs to him that Death is kind of a silly thing to be having existential debates with. Zing! Death can't take a joke, so the Wanderer goes on to the Light.

Doofy as it was—and geez, c'mon, I was 17—the story was very true to the spirit of its time. I submitted it to the *Lane Tech Prep* (my high school's literary magazine) and it was accepted, and published in the winter 1970 issue. As I did with a lot of my short material at that time, I passed it around among my friends, who were much impressed. As the years went by and we kept in touch, we sometimes reminisced and giggled over the story, knowing that hey, in 1970, it was cool. And as with so many other things, I (and my friends) grew out of that silly kind of cool.

I did a goofy thing in the spring of 2013: I wrote a sequel. "Doing Doughnuts on the Edge of Darkness" is very much me making fun of my younger self and the stories I had written while hoping to become a literary sensation. I dedicated it to two of my friends whom I had known from grade school, who giggled with me at the silliness that we had all been prone to fifty years ago. Hey, maybe Death isn't a cosmic force, or some dark archetype. Maybe Death is just another guy punching a clock and doing his job. Maybe the poor guy needs a promotion—or at least a long vacation, and, hell, a girlfriend.

As another friend told me once, decades ago: "You'd redeem the devil himself if you had the chance!"

Damn straight.

Dialog on the Edge of Darkness

From the *Lane Tech Prep*, Winter, 1970

The Wanderer and Death walked side by side on the shore of the sea. Its waters clapped morosely on the silent black sand as they passed and stretched beyond sight into the unfathomable void. The water was blacker still than the sand they trod. The light was not from the sun, for there the sun had never shone. There was only the bleakness of black sand and blacker sea, and everywhere the mumbling of pending oblivion.

The Wanderer, hands clasped behind him, listened with eyes downcast to the words of the creature walking beside him.

"And that is hardly a reason to refrain from embracing me. Not at all. I am more human than even you, for many times more than many can be, have I taken humans like you into myself. They have taken me gladly—we are one."

"Then you must be more animal than human," the Wanderer chanced. "I have heard that animals will accept death willingly."

Death shook his head. "Animals do not know me. They cannot know me. Life is not given to them; it is thrust upon them. Likewise is their death thrust upon them. They have no choice in the matter."

Death's lopes quickened, for the Wanderer had begun to walk more briskly. His hands wrung nervously behind his waist, and his forehead was wrinkled with quiet and intense thought.

"I have loved. I have loved three times. I cannot embrace you."

Death's teeth glinted crookedly in his ghostly jaw.

"If that were a reason, I would not exist. All humans have loved as surely as they have lived. But only selfless love can cower Death. Only selfless love, and love which is not intermingled with hate at some time in its duration. You do not seem to understand that love and hate are closely akin—one and the other are easily interchanged. One can fade into the other in the least instant." Death's awful face looked into the Wanderer's with triumph. "Did that not happen to your first love?"

The Wanderer nodded sadly.

"Ah, it was love," Death agreed, "but it was selfish love. You desired her only for your own gratification. You had no thought for her. You had not yet learned of selflessness. And in the end it turned her against you. She hates you."

"That is true," the Wanderer accorded. "And I have regretted it for decades. But my second love..."

"...was a selfless love," Death completed with a discolored leer. "Selfless in the extreme. You gave her everything, and got nothing. You loved her, and were rewarded with indifference. Then your love turned to hate. You hated her."

Again, the Wanderer nodded sadly. Death glimmered with a loathsome light.

"Your third love was perfect, was it not? Yes, you have been thinking that all along. You gave yourself unselfishly to her, and she likewise to you. You walked side by side in life, as we walk side by side here now. You raised your children well. Does this not constitute a perfect love? Aye, a most perfect love. But you forgot the final moments of life. When the driving machine malfunctioned fatally, what did you reach for? Her? Or..."

"...the door. I reached for the door." The Wanderer brought his hands to his sides, and clenched his fists for a moment. "I reached for the door."

Death raised a rotten, decaying finger. "You reached for the door because you loved yourself more than you loved her. And she, though you could not have noticed, did the same. In her mind were only thoughts of escape. All who have lived would have done the same. All reach for safety before they reach for the one they love. It is indelibly part of the human mind. Self-preservation is a product of the self-love. It is said you can love no one more perfectly than you can love yourself."

Traces of puzzled anger passed over the Wanderer's face. "But that would essentially condemn the entire human race!" Death nodded his shadowy head. "Did I not say that human love cannot refute Death?

Then the only sound was the dismal slapping of the ceaseless waves upon the stagnant shore. The Wanderer fell silent, his entire life passing before the eyes of his mind. He was human, but death was immeasurably more human than he. He had loved, but he had also hated; in the end he had found that he had loved only himself. What else had he done?

What could any man do but live and try to take all of life's complications in his stride, for good or for bad?

The Wanderer raised his hand, wiped the grime of fifty years' life from his cheek, and looked foresquare at the creature stumbling along beside him. Was Death really so terrible an entity, infallible and all-powerful? Was he all-truthful? Or was he anything more than a Halloween spirit; a grasping, unscrupulous bag of old bones and rotting cloth? Like anything he had ever encountered, was Death not a hurdle to be challenged, and to be won? Surely Death could not be an end in himself, to be embraced gladly—any more than a minor pain or a fatal accident. One could grit one's teeth against pain, or perhaps foresee an accident, but what could one do about Death? There had to be a common denominator somewhere, something which could be set against anything, including pain, accidents, and even Death. Then, in a flash of quiet revelation, he found that there was, and that he had known and used it all his life.

"I have laughed," the Wanderer said with a weary grin.

Death halted his heaving step, and stood crookedly still upon the featureless beach. His eyes glowed with a surging green lava, and slowly his shadowy head began to nod

"Aye, you have laughed. If nothing else, Wanderer, you have laughed. And now you have won." Death raised his eternally crumbling hand in a final salute, not hatefully or spitefully, but in deference to the man standing before him. "You have placed yourself above your little troubles with your laughter, and in your realization that I am as human as they, you have placed yourself above me. I take your leave."

Death turned, still crookedly, and strode off into the darkness from which they had come. The Wanderer did not look after him. Far ahead, a new wind was beginning to blow, casting aside the darkness, revealing a sea of sparking blue and sand of glittering gold.

His wandering was over, and it looked like a beautiful eternity.

Doing Doughnuts on the Edge of Darkness

Published here for the first time ever

The Wanderer and Death walked side by side on the shore of the Sea. Its waters clapped morosely on the silent black sand, interrupted only by the occasional beer can and squeeze bottle of SPF-24. The light was not of the Sun, for this was not exactly Miami Beach.

"I have laughed," the Wanderer said.

Death twitched. "Shit! *You* again!"

The Wanderer bumped Death's shoulder with his fist. "Hey, man, good to see you!"

Death continued to face forward. "I have nothing to say to you."

"I have learned where God parks his dune buggy," the Wanderer said.

Death pulled out a stained leather-bound book and thumbed through it. "That's not a valid code. Why don't you just say, 'This has been willed where what is willed must be' so I can drop-kick you into Purgatory and be done with it?"

The Wanderer shook his head. "Doesn't work anymore. God flashed the firmware or something. Besides, I've already spent time in Purgatory. It looks a lot like Hoffman Estates."

"Who told you that!"

"I grew up in Hoffman Estates."

"No, about the codes."

"The chick who gave me the treasure map. I tried to ask her out but she only dates guys with wuss names like 'Virgil.' Great bod, though."

"You met Beatrice and all you did was look at her body."

"I met her because somebody pulled strings. Your high school classmate Ti. She's now the Muse of Twitter. You can follow her at @adhdmuse."

Neither spoke for a long time. Death side-stepped around a broken Dos Equis bottle. His voice was wistful. "So Ti made it to Muse. I'm surprised. She couldn't even finish her own sentences."

"She had a crush on you. She sat next to you in AP Metaphysics. You used to give her the right answers."

Death looked down at the sand. "Well, what I thought might have been the right answers."

"Hey, it's the thought that counts. Did you two ever get it on?" Silence. The Wanderer gripped Death's shoulder and shook it a little. "C'mon! You can talk here. We're both dead. Did you and Ti ever do it?"

"You're dead. I'm Death. I made it that far, at least. Don't get cocky."

"You're dodging. Did you ever get into her pants?"

"Look. I don't even have skin, much less a…"

"Dodging!"

Death sighed. "I tried to kiss her sophomore year. She ducked."

The Wanderer laughed, and clapped Death on the back. "I would too! But you *tried*."

Death nodded. "Ti was ambitious for a ditz. She wanted something better. I took the ZATs and applied to zombie school, but I didn't have the brains. I never got the knack of shambling, either. She CLEPed out of harpy school and got her White Lady certification in less than a year. All the other girls in my class took their banshee degrees and quit. Now they just drift around all night, screaming their heads off and daydreaming about Sean Connery."

"Ti told Beatrice that you were a pretty nice guy, for Death."

"Right. And you know how nice guys never get…lordy, *why* am I having this conversation with you!"

"You have loved," said the Wanderer.

"Stop that!"

The Wanderer clapped Death on the back again. "I think you're bored. And I think Ti was right." He pulled a sheet of paper from his shirt pocket and held it up to his face. He then squinted into the murky distance. "Those two dunes up ahead look familiar. I think we're getting close."

"To what?"

"To having the first good time you've ever had in your whole…death. Life. Whatever."

A few minutes later, after threading their way around several small dunes, they found themselves standing in front of a battered, dark-green VW chassis with fat tires but no roof. From a fiberglass rod on the rear bumper flew a small pennant bearing the Tetragrammaton: יהוה

The Wanderer jumped over the driver's door without opening it, then leaned across and opened the passenger door. "Get in!"

Death hesitated, then shrugged and got into the passenger seat. He started poking around in the cushion cracks with bony fingers. "There's no seat belt."

"What are you afraid of? Dying?"

The keys were in the ignition. The Wanderer wiggled the stick shift and released the parking brake.

Death looked over his shoulder and stared at the pennant. "We're stealing God's dune buggy!"

The Wanderer twisted the key. The engine roared to life. "C'mon. He made the whole damn universe in six days. How long do you think it'll take Him to make another dune buggy?"

"But He's…God!"

"He'll forgive us. It's what He does for a living."

The Wanderer threw the stick into first and took his foot off the clutch. Spitting black sand, the buggy leapt forward, throwing Death back against the cushions. The Wanderer stomped on the accelerator, threw his head back, and cheered.

"Yeeeeeeeeeeee *HAH!*"

The buggy hit a small hummock and got ten feet of air. Then the dunes were behind them. The Wanderer swung the wheel hard left, and then hard right.

"Handles great! God knows his shit with transmissions!"

"You're crazy!" Death leaned forward and gripped the top edge of the windshield with both bony hands.

"Damn right!" The Wanderer pulled the buggy into one tight circle after another. He glanced to his right, and punched Death's shoulder again. "Relax! Your knuckles are white!"

"My knuckles are always white!"

"You must not get out much. Hey, there's some bumps up there. More air!"

The Wanderer floored the buggy for several long seconds. The vehicle bucked high and almost nosed into the sand. It recovered after several bounces, but both he and Death got a faceful.

Death took one hand away from the windshield to brush sand off his hooded robe. "You wound up here the first time fooling with fast cars. You don't learn, do you?"

The Wanderer spat sand and leaned into the tightest turn yet. "Second time too! I bought a beat-up '75 Vette and put crappy tires on it. Flipped it three times and wrapped it around a bridge pillar on the Dan Ryan. I was only 23, but man, what a ride!"

The Wanderer backed off on the gas and drove a straight line along the beach. He reached into his shirt and pulled out a fifth of Jaegermeister. He tossed back a long swallow. "Something to wash down the sand!" He waved the bottle in Death's direction.

Death took it. He thrust the neck of the bottle between his bleached jaws. Jaegermeister dribbled out of his neck. The Wanderer swung the buggy into another doughnut and laughed wildly.

Death threw the empty fifth out into the darkness and laughed as well.

"I see a hill!" The Wanderer floored the buggy again. For fifteen seconds the engine roared. Then they crested the hill and took air for thirty feet. The landing was rough. Death was nearly thrown out of the buggy entirely. They both heard something snap.

"Lost a ball joint!" The buggy veered hard right, dragging its bad wheel, bounced where the sand grew wet, and went into the sea. The buggy nosed down and stopped, with dark water up to the level of the front seats. The Wanderer sloshed the water with one hand. "It's warm!" With both hands he splashed water in Death's face. Death tried to do the same, but his skinless hands didn't catch much water.

The Wanderer and Death laughed so hard that Death's bent-over skull touched the dashboard.

220

The Wanderer stood in front of the driver's seat, knee-deep in water. "I have laughed!" he exulted, then reached down and dragged Death to his feet by one elbow. "And so have you! It's over! Now we can both reincarnate and do it all again!"

"But…I'm Death!"

The Wanderer squeezed Death's sopping left sleeve. "You've been doing this shit long enough. Let somebody else have the job. You can be my guardian angel." The two regarded one another until they both began giggling, and then broke again into gales of laughter. "Ok. Maybe my imaginary friend. Or the monster in the closet! I'll let you have the bottom bunk if you want it. Just don't scare my little sister, or we'll both be in deep shit. Deal?"

"You're crazy!"

"Deal?"

Death looked out across the darkness to the shadowed horizon. Rays of brilliant light were breaking through, scattering the clouds and turning the water luminous blue.

"I belong here."

"Like hell. We'll get you out of that ratty bathrobe and into some khakis. If you can get word to Ti somehow, we could double. Deal?"

Death shook his head and laughed lightly. "Deal."

"Cool! In the meantime, skinny dip!" The Wanderer ditched his shirt and pants and jumped over the driver's door, throwing a huge curtain of water into the buggy that soaked whatever dry spots remained on Death's robe.

"C'mon! The water's fine!"

Death considered. "Do bones float?"

"It's four feet deep, you chicken! Come freaking *on!*"

Death released the belt of his robe and wriggled out of the sodden cloth. He stood up on the seat and looked apprehensively at the water.

"We are *so* grounded," Death said, and jumped.

The Parable of the White Tile

Published here for the first time ever

Centuries ago, during the Age of the Great Cathedrals, a mighty church was rising against the rolling green hills of a distant Christian land. The king of that country had retained the world's greatest architects, masons, sculptors, and artists to build the church, which would be a task of many years, perhaps decades. The greatest of all the artists that the king had employed had come from far away, and made his home in the shadow of the church, knowing that he would be pouring most—and perhaps all—of the rest of his life into its completion.

He was an artist of a special skill, the creation of mosaics. With nothing but colored tiles he could paint scenes and landscapes so real, so luminous, that they seemed to have a life of their own, as though they were windows into the ineffable realms of Heaven itself. His task in the building of the church was a mosaic above the main altar, sixty feet high, depicting Mary, Queen of All Saints.

The mosaic would require tens of thousands of colored tiles. The Artist made each of the tiles himself, alone, by hand, at a small bench behind the main altar. Each tile was precisely what the mosaic required. Each one was shaped individually in the Artist's hands, and no two were alike. His skill was great: No more than were needed were made, none were ruined, and none were thrown away. After the tiles had been colored and fired, the Artist took them up on the scaffold himself, and cemented each tile individually and precisely into its place in the great mosaic.

The Artist was the greatest that his craft had ever produced, and he had promised Jesus and Mary that this mosaic would be his masterwork. God saw how the Artist loved the tiles he had crafted, just as God loves all of His children, and in a special way God allowed the tiles lives according to their natures, and made them recognize the Artist as their master, because tiles have neither minds nor souls with which to recognize God. The Artist spoke to the tiles as he shaped them, fired them, painted them, and positioned them in the mosaic. As the years went on and the mosaic took shape, the tiles would speak to one another and to the Artist, who praised each of them for its part in the greater work that was unfolding. The tiles listened to the Artist, and they were happy.

All but one. At a particular place within the mosaic was a white tile. The tile knew the tiles all around it, but no more than that, because a tile within a mosaic cannot see the picture of which it is a part. The white tile looked to its neighbor tiles, and realized that all of them were made of gold. The white tile was large, and its angles were irregular. All of the surrounding tiles were smaller than the white tile, and of compact and regular shapes. Where the golden tiles had neat corners, the white tile had sharp spikes. This made the white tile unhappy.

One day, as the Artist was positioning new tiles into their places in the mosaic, he heard the white tile calling out to him: "Master! Master! Why am I so strange and ugly?"

The Artist heard the white tile, and stepped down a few rungs on the scaffold to where the white tile was, so that he could speak to it: "You are not strange and ugly, my child. You are precisely what I needed you to be."

The white tile was not convinced. "But all the tiles around me are made of beautiful gold! I have no color at all!"

The Artist shook his head, and smiled. "White is the greatest of all colors, dear one, because it contains all other colors. Every color of the rainbow lies within you."

The white tile was still not happy. "But I am huge and gross, and have no shape. All of my angles are sharp, and nothing about me is regular. I am nothing but a jagged, ugly, spiky white blotch. Why, master? *Why?*"

The Artist leaned forward toward the tile. As his eyes grew closer to the mosaic, he reminded himself that he could see the tiles the way they could not see themselves, and that he could understand their places in the heavenly image in a way that none of them could ever understand. So it was with great tenderness that he reached out a fingertip, and gently touched the white tile while he replied:

"Because, my dear child, you are the dazzle at the center of Our Lady's golden crown."

Creation

Published here for the first time ever

It was always Always o'clock at God's house. God's house was the only house on a very short but very wide street with only one address, which was Everywhere. The house was bigger than you or I might need, but He'd been very comfortable there since, well, Always.

God was looking at the clock over His kitchen sink late one Always, and puffed out His cheeks. "Too quiet in here!" He said, and He *meant* it.

So God went into the livingroom, which looked out on Everywhere, and sat down at His piano. He played a jazzy number that always got Him into a good mood. This time, though, He decided to improvise a little, and the tune changed. The tempo went up, and God found Himself tapping His feet. This was worth more than just a few bars for practice's sake. "Hold that thought!" he told the piano, which kept on playing as God got up, and He danced.

He danced in place for a moment as the tune played, but that wasn't good enough this time. No, this was a tune worth something more, something new, something that even God hadn't done in all the Always that he'd been living there at Everywhere.

So He danced out the front door out onto the lawn. "It's been Always for way too long!" God said. "I want time, I want days, I want motion, I want *change!*" He raised His arms high, and in a voice that shook the (very large) rafters of his (very large) house, God shouted:

"Let…There…Be…Light!"

And with an enormous noise (maybe the biggest bang that ever was) there was light, and there was shadow, and there were dazzling spotlight colors on the stage where God's wild dance went on.

Suddenly it was no longer Always. Something had changed. Things were happening. A glance through the kitchen window showed that the hands on His clock had started to move. The first day was underway, and things were turning out very well indeed.

So God danced in the light for a moment, as the tune danced with Him, and the light with the tune. He looked up at the empty skies, and said, "Hey, I can do better than *that!*" So in a voice that made the light tremble, He shouted:

"Let…there…be…a…firmament!"

When God raised His arms again, stars and comets and galaxies streamed out of His sleeves and poured up into the skies. God spun in a huge circle, and the light spun out from Him, filling the empty heavens with lights like even He had never seen before, though (as He would admit if pressed) He had painted once in acrylics. The greatest light He called the Sun, and when the Sun looked bright (but lonely) He spun out another light called Moon to be Sun's dance partner, and Sun and Moon danced in a wild reel as the second of God's (very) long days came to a close.

"Now that's cool!" God said, seeing the greater and lesser and all very beautiful lights that came from His dance and now danced with one another. "But I'm just warming up!"

"Let…there… be…a…world!" God shouted.

So God danced down the short (but wide) street that was Everywhere, and raised His arms again. From His sleeves poured water, stone, sand, and mountains, hills and valleys, rocks and boulders, which spread themselves out as He spun in His dance, into islands and continents, bays and rivers, peninsulas and isthmuses (which are easier for God to spell than for us), water beside land, and land beside water, all dancing to the tune that now rang from the walls of canyons and echoed through caves and dells. And before you know it, another day had passed, but God was on a roll, and as long as His days were He was nowhere *near* done yet.

No way. The firmament was good. The land was good. The seas were good. But for all the wonder of the dancing universe, there was something missing. But what? Ha! Just the thing! God looked out at the dark waters, and, snapping his fingers, shouted again:

"Let…there…be…LIFE!"

Sparks came from His fingers as they snapped, and those sparks rode out over the empty waves like fiery strokes of lightning. Into the rolling water they settled, where the sparks joined with the water and the sand, grew, and danced. They were tiny things that soon became larger things, things that darted and swam and glowed, all with the spark that came from God's hand and God's hand alone. Fish and seaweed, worms and squid and lobsters, all joined God's dance amidst the waters, sunlight dancing with moonlight, as the fourth day ended.

The life in the waters was good; better, in fact, than God had hoped. But the land remained empty, and the next step was pretty obvious. God danced beneath the light down to the shores of the teeming waters, and gestured to the sparks that danced there with Him:

"Let…life…find…the…land!"

And from the oceans and bays and rivers came the small things that swam and slithered and crept, up from the shores onto the land. And so that they could better join God's dance, they changed. They grew legs and wings, and by the spark that was in them they ran and flew, dug and climbed, and they filled the world so that no place was without the spark that God had struck from His fingers, all in the rhythm of His dance. God's new universe resounded with rhythm as Day 5 came to a close.

All was good; the light, the skies, the seas, the land, and the life that had sprung from the rhythm in God's fingers. But God realized that even though the whole universe danced, He still danced alone, because there was nothing in the universe even a *little* bit like Him. Water had land, Sun had Moon, life had life in all its countless forms.

But who would dance with God?

This was a difficult question. Who indeed? There had always been Always here at Everywhere, but now there were days and time and a universe and a dance. Everything had a partner…all but God. So God picked up a handful of wet dirt from the shore where land and water met, and saw that the sparks of life were there in His hand, shining in the dance but still no fit partner for God.

God set the handful of dirt on the ground. He rolled up His great sleeves and took a deep breath, for this was a task unlike any other that He had ever attempted:

"Let…there…be…HUMANITY! And let them be made in My image and likeness, that they may join My hand in their hands and dance this dance with Me on this day and whatever days may yet come!"

And God breathed His life into the moist earth before Him, and it was Humanity, formed in God's image and likeness. Humanity breathed the air, and thought, and stumbled, whole but still somehow incomplete. So God filled Humanity with the mystery of love that comes from God and is God. Humanity became many

Someones bound together and to God in love and friendship, so that every Some-one would always have another Someone as a partner in God's dance.

Yet even then something bothered God, which is not an easy thing to do. He had held something back, something potent and God-like and maybe even a little bit scary. He didn't want pets. He wanted partners, companions, or even… deep breath here…co-creators within this new Everywhere, and the Always that now ticked with time and change and unfolding wonder. So God took a seventh breath, and with only a little catch to that great breath made it so:

"Let…Humanity…be…FREE!"

And like God they were free. In their freedom Humanity began to improvise, dancing a new dance that echoed God's dance but was not God's dance. They danced away from Him, out into the wide universe that He had made.

God was lonely again, but He was patient. He had done what He had meant to do, and He knew what He had done, and it had taken a bit of the wind out of Him. So He kicked His shoes off and heaved a little sigh, and spent the seventh long day looking out at His new world. The fish, the animals, and the birds all knew Him from their birth. But Humanity, well…Humanity was *different*. Free-dom was a funny business. Freedom meant mistakes. (At least if you weren't God.) Mistakes *hurt*. God went back into His house, and as He passed by the piano, rattled out a few bars of a (slightly) sad song. It had only been a day, but God's days were long, and he *missed* them.

"Well, hey," God said to Himself. "It took me seven long days to get it all done. Let's give them some time to learn the way back." God took out some corn chips and salsa, lit the grill, and put a few sodas in the fridge. He dug out the Band-Aids and the Bactine, and a family-size bottle of Tylenol. He looked at the clock, and glanced at the big front door to make sure it was open and the welcome mat was out.

It was. And to balance the scariness of freedom, God made Humanity a prom-ise, and threw a rainbow up in the sky over Everywhere as His seal, while He spoke the words: "There is a Someone for Everyone, and a place in Everywhere for Everybody. So I have made it, and so it will be, *ALWAYS!*"

It was half-past eight. Could that be music in the distance? And dancing?

"Here they come!" God shouted, and opened His mighty arms to welcome His children home.

About the Author

Jeff Duntemann has been published professionally since 1974, in both science fiction and technical nonfiction. His stories have appeared in *Isaac Asimov's Science Fiction Magazine*, *Omni*, the *Orbit* and *Nova* anthology series, and several standalone print anthologies. Two of his short stories have appeared on the final Hugo Awards ballot. His first hard SF novel, *The Cunning Blood*, appeared in hardcover in 2005 and as a Kindle ebook in 2015. His fiction may be characterized as "Human Wave," a term coined by author Sarah Hoyt to indicate generally upbeat tales that place story ahead of message, and affirm rather than denigrate the human spirit.

On the nonfiction side, he has worked as a technical editor for Ziff-Davis Publishing and Borland International, launched and edited two print magazines for programmers, and has numerous technical books to his credit, including the bestselling *Assembly Language Step By Step*. He wrote the "Structured Programming" column in *Dr. Dobb's Journal* for four years, and published dozens of technical articles in many magazines. With fellow writer Keith Weiskamp, Jeff launched The Coriolis Group in 1989, which went on to become Arizona's largest book publisher by 1998.

After retiring from technical publishing in 2009, Jeff created Copperwood Press for new and reprint publications in several areas from history to SF and fantasy. Outside of writing and publishing, Jeff's interests include programming, electronics, amateur radio (callsign K7JPD), astronomy, telescopes, history, psychology, and kites. Jeff lives in Scottsdale, Arizona with his wife Carol and three bichon frise dogs.

Read Jeff's blog Contrapositive Diary:

www.contrapositivediary.com

and his tech projects site, Jeff Duntemann's Junkbox:

www.junkbox.com

Follow Jeff on Twitter @JeffDuntemann

Hard SF Action-Adventure at Its Best

Framed for murder by Eath's world govern-ment, Peter Novilio is offered his freedom in exchange for a reconnaisance mission to the surface of Hell, Earth's escape-proof prison planet. Hell is infected with a nanobug that eats electrical conductors, making compu-tation and spaceflight impossible. There is a way back, known only to his grim mission partner, Gayle Shreve.

But Peter has a secret too: In his bloodstream he carries the Sangruse Device, an outlawed nanotech AI of fearsome power, with its own reasons for visiting Hell. Peter soon realizes that he is a pawn in a covert war among Earth, Hell's ingenious inmates, and the deadly mechanism in his veins. For as fearsome as it is, the Sangruse Device itself is afraid—and the fates of whole worlds would depend on the threat that the Cunning Blood had discov-ered outside of space and time.

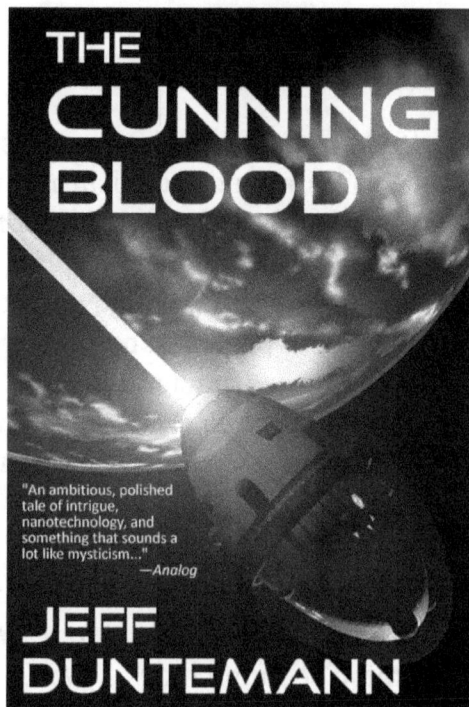

THE CUNNING BLOOD

"An ambitious, polished tale of intrigue, nanotechnology, and something that sounds a lot like mysticism..."
—*Analog*

JEFF DUNTEMANN

Kindle EBook $3.99   Trade Paperback $12.99

"[Jeff Duntemann] returns with an ambitious, polished tale of intrigue, nanotech-nology, and something that sounds a lot like mysticism...This one has a decent chance of ending up on award ballots." —Tom Easton, *Analog*

"The book is absolutely *au courant*, and actually extends the Great Work of SF in several unexpected directions. Like most ambitiously sprawling *sui generis* books, this one delivers the sense—as with the work of the recently departed Charles Harness—that the author has chucked every idea he had during the writing of the novel into the pot."   —Paul Di Filippo, *Science Fiction Weekly*

"Whether your interest is in scientific ideas, widescreen action, or sheer flights of imagination, you will find much to enjoy in *The Cunning Blood*. I look forward eagerly to Duntemann's future work." —David Hebblethwaite, *SFSite*

For More SF in the Grand Tradition...

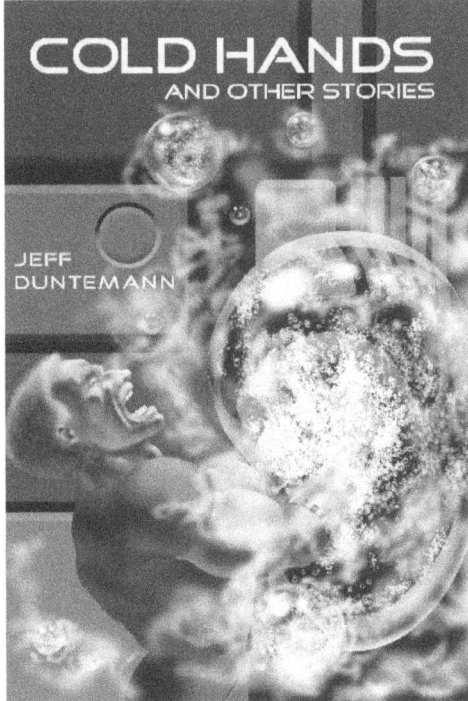

Magic & Monsters Meet Software & AIs

Having cheated a powerful magician out of ten nuggets of pure magic in a rigged card game, Bartholomew Stypek needs a place to hide. As a spellbender he is a partial magician who can read and change magic spells, but, absent a stash of magical force, cannot cast his own. With his familiar spirit Pickles and the ill-won Opportunities, Stypek leaps blindly across universes, hoping to be dropped someplace far away and without magic...and lands in the break room of a small advertising agency in Upstate New York.

Because our universe doesn't support spirits, Pickles manifests as the local equivalent: AI software in the agency's heavily networked copier. She wanders into a nearby corporate network looking for allies, and discovers a virtual universe where AIs live and train for jobs as AI products. Pickles is soon seducing Simple Simon, a naive AI created to control a huge robotic assembly line. Stypek, mistaken for a penniless CS intern, is taken in by the ad agency's copywriter. Expecting the usual suspicion and contempt, he is humbled by the kindness he's shown, and one by one uses the stolen Opportunities to help his new friends with their problems.

But Jrikk the magician isn't so easily thwarted. Soon Stypek, Pickles, Simple Simon, and their human and virtual friends must fight for their lives against the evil force sent to retrieve Stypek to the magician's dungeons.

Kindle EBook $3.99   Trade Paperback $12.99

"*Ten Gentle Opportunities* represents the best that science fiction and fantasy have to offer. It blends the two genres in a clever and original way. It presents near future tech that is plausible, delightful, and a little scary. Best of all, it provides an exuberant and unapologetic adventure that incorporates action, violence, romance, and robots in ways that are both exciting, fun to read, and even a little bit educational."

—Jon Mollison, Seagull Rising

Meet Larry: He's Your Worst Nightmare's Worst Nightmare

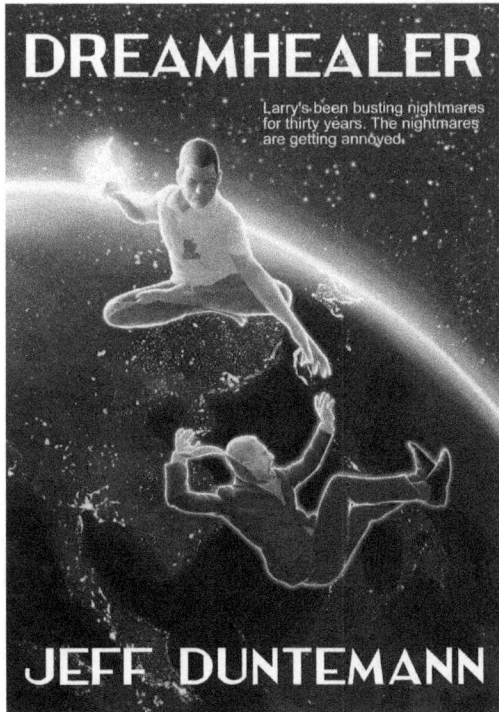

Two Short Novels of the Drumlins World
In One Volume

Step up to the pillars in front of the big bowl of gray dust, tap the pillars 256 times in any pattern, and *something* appears in the bowl. What? *Almost anything.* This is the Drumlins World, an alien planet where the aliens have gone, leaving their replicator machines behind, still working, still capable of producing 10^{77} different things—one for every atom in the universe! Here are two short novels by Drumlins creator Jeff Duntemann and *Brass and Steel* author Jim Strickland:

• In *Drumlin Circus*, a circus master attempts to free his wife from the shadowy Bit-space Institute, with the help of his bodyguard clowns (one a former Institute man with a grudge against the organization) and a preteen girl with a whistle drumlin that can make other drumlins, animals, and even human beings obey her. Intrigue, steam, smilodons, airships, and nonstop action!

• *On Gossamer Wings* tells the story of a teen girl who cannot speak, but has a strange talent allowing her to find exactly the drumlin she wants among the 10^{77} possibilities in the Thingmakers. Her farm town neighbors hold her in contempt, and she struggles to realize her dream of building a drumlin flying machine before the Bitspace Institute fully comprehends the breadth of her powers.

Kindle EBook $2.99 Trade Paperback $12.99